Praise for *What It's*

"Groundbreaking research that shows that dog emotions are similar to people's. Training dogs to voluntarily lie still in the MRI brain scanner was a brilliant way to explore the workings of their brains. Dog lovers and neuroscientists should both read this important book."

Temple Grandin, author of *Animals in Translation* and *Animals Make us Human*

"We know a lot about the intelligence of animals and nearly nothing about their brains. Greg Berns is changing all of this by means of non-invasive techniques that respect the animals. He is boldly going where no one has gone before, offering a lively, eye-opening peek into his neuroscience kitchen."

Frans de Waal, author of *Are We Smart Enough to Know How Smart Animals Are?*

"Berns has done it again; woven a compelling story with a scientific revolution. From building an MRI simulator in his living room to tracking down one of the four remaining brains of the extinct Tasmanian tiger, Berns takes us on an incredible journey of exploration and discovery. Marvelously written and intellectually engaging, *What It's Like to Be a Dog* will establish Berns as one of the most skilled neuroscientists of our day, as well as someone with the intuition that understanding other animals will lead to greater insight and knowledge about ourselves."

Brian Hare, bestselling author of *The Genius of Dogs*

"Have you ever wanted to peek inside the mind of a dog? Gregory Berns' brain scanner does precisely that. But this book also contains many remarkable insights into the inner lives of other animals. Dolphins, sea lions, raccoons, Tasmanian devils—even the long-extinct Tasmanian tiger—they're all here. A fascinating journey towards an understanding of what dogs—and their mammalian cousins—might be thinking about us."

John Bradshaw, author of the *New York Times*
bestsellers *Dog Sense* and *Cat Sense*

"It's the rare neuroscientist who has the patience and curiosity to train dogs to hop into an MRI machine, tails wagging. Or delve into the mysteries of the dolphin brain. Or venture to the far side of the globe to find the brain of an extinct, yet still fascinating species: the thylacine. Thankfully, Gregory Berns did all of these things. In this big-hearted book, he applies cutting-edge science to questions that have never been so timely: How do other animals perceive their worlds? How do they experience emotions? How does their language work? *What It's Like to Be a Dog* is a delightful, illuminating look at the minds and lives of our fellow creatures."

Susan Casey, author of *Voices in the Ocean*

"Berns mixes personal stories of dogs and dog lovers with elegant scientific experiments that show the surprising complexity behind many canine daily behaviours: a fun, fascinating and illuminating read."

New Scientist

"One of the most delightful things about *What It's Like to Be a Dog* is the attention Berns pays to each dog's individual quirks."

New Yorker

WHAT IT'S LIKE TO BE A DOG

Also by Gregory Berns

How Dogs Love Us: A Neuroscientist and
His Dog Decode the Canine Brain

Iconoclast: A Neuroscientist Reveals How to Think Differently

Satisfaction: The Science of Finding True Fulfillment

WHAT IT'S LIKE TO BE A DOG

AND OTHER ADVENTURES IN ANIMAL NEUROSCIENCE

GREGORY BERNS

ONEWORLD

A Oneworld Book

First published in Great Britain, the Republic of Ireland and Australia
by Oneworld Publications, 2018
This paperback edition published 2019

Published by arrangement with Basic Books, an imprint of
Perseus Books, LLC, a subsidiary of Hachette Book Group, New York,
New York, USA. All rights reserved.

Copyright © Gregory Berns 2017

The moral right of Gregory Berns to be identified as the Author of this
work has been asserted by him in accordance with the Copyright,
Designs, and Patents Act 1988

All rights reserved
Copyright under Berne Convention
A CIP record for this title is available from the British Library

ISBN 978-1-78607-489-8
eISBN 978-1-78607-362-4

Typeset by Jack Lenzo
Printed and bound in Great Britain by Clays Ltd, Elcograf S.p.A.

Oneworld Publications
10 Bloomsbury Street
London WC1B 3SR
England

Stay up to date with the latest books,
special offers, and exclusive content from
Oneworld with our newsletter

Sign up on our website
oneworld-publications.com

For Callie

Falkirk Community Trust	
30124 03108463 7	
Askews & Holts	
591.513	£9.99
MK	

Contents

	Introduction	1
1	What It's Like to Be a Dog	9
2	The Marshmallow Test	27
3	Why a Brain?	49
4	Seizing Sea Lions	75
5	Rudiments	101
6	Painting with Sound	117
7	Buridan's Ass	137
8	Talk to the Animals	159
9	A Death in Tasmania	183
10	Lonesome Tiger	211
11	Dog Lab	235
	Epilogue: The Brain Ark	257
	Acknowledgments	261
	Notes	265
	Index	283

Your dog is a philosopher who judges by the rule of knowing or not knowing; and philosophy, whether in man or beast, is the parent of gentleness.

—SOCRATES (PLATO, *THE REPUBLIC*)

Introduction

Until bin Laden was killed, I hadn't given much thought to animal minds.

It wasn't bin Laden but Cairo, the dog on the mission, who caught my attention. Cairo was a military working dog who could do amazing things, like jump out of helicopters. His ability to tolerate noisy, chaotic environments gave me an idea that, in retrospect, seemed so obvious that it was strange no one had thought of it before: *If dogs could be trained to jump out of helicopters, then surely they could be trained to go into an MRI scanner.* And why would I want to do that? To figure out what dogs were thinking, of course.

The timing was serendipitous. I had spent thirty years in science, first training to become a bioengineer, then a physician, and eventually specializing in the use of MRI to study how the human brain makes decisions. My favorite dog—a pug named Newton—had died the previous year, and somewhere in the back of my mind I had been pondering what the dog-human relationship meant. Had Newton loved me in the same way I had loved him? Or had it all been a sham, an innocent duplicity propagated by dogs to act all cute and stuff in exchange for food and shelter?

In Newton's place, our family had adopted a skinny black terrier mix, whom we named Callie. She was as opposite a pug in
appearance as in demeanor. Insecure and high-strung, she tended
toward bullying the other dog in the house, a sweet golden retriever who didn't put up any resistance. But besides Callie's feisti-
ness, she had another trait that no other dog I had owned had
possessed: curiosity. Callie loved to learn new things. The usual
dog tricks were a breeze, and Callie soon discovered interesting
things about life in a human household. Useful tidbits like how
door handles worked. No need to wait for the humans to get into
the pantry. Callie figured out that if she rose up on her hind legs,
she could use her front paws to pull the handle down and out.
She did it with such alacrity that you would have thought she was
a capuchin monkey with opposable thumbs. This skill, unfortunately, was a hard-earned one that landed her in the ER with a
stomach filled to the brim with who-knew-what.

I needed to give Callie something to do. Why not put her
skills to more productive use than finding creative ways to swipe
food, such as training her to go into an MRI so I could figure
out what she was really thinking?

I sought out the help of Mark Spivak, who ran a local
dog-training company called Comprehensive Pet Therapy. Mark
was up for the challenge, and we began working through all
the little details necessary to train Callie to hold still inside an
MRI long enough for us to see how her brain worked. Sedation
was out of the question for two reasons. She would have to be
completely awake so that we could see how her brain processed
things like smells, sounds, and, most importantly, communication from her owner—me. And because we set out to treat her
in the same way that we would treat a human participating in
an MRI study, she had to be able to leave the scanner whenever

she wanted. Like a human, our canine participant would be a volunteer. That meant no restraints.

I built an MRI simulator, which I parked in our living room. We constructed mockups of the "head-coil" that picked up signals from the brain, and Mark and I quickly taught Callie how to shimmy into it. Although the process was mostly trial-and-error, and we hit many speed bumps along the way, it turned out not to be as difficult as we initially thought it would be. With just a few months of training, Callie had graduated from a discarded dog in an animal shelter to the first dog to have her brain scanned voluntarily while fully awake and unrestrained.

Encouraged by our success, we solicited the help of the local dog community to join this groundbreaking project to investigate the workings of the canine mind. Much to my surprise, we had no shortage of volunteers. So many volunteered that Mark and I developed a tryout protocol so we could identify the dogs who would be most likely to succeed in MRI training. Within a year of Callie's first scan, the team had grown to nearly twenty dogs. To accommodate all these dogs and people, we held MRI practice every Sunday afternoon, alternating weeks between the "A-Team" and "Bravo Company."

We began with very simple experiments to see how dogs' brains responded to hand signals that cued the delivery of treats. In humans, it was already known that a key brain structure, called the caudate nucleus, responded in anticipation to things that people liked, such as food, money, and music. So when we discovered that the dog's caudate reacted similarly to hand signals, in anticipation of the treats, we knew we were on to something important. The dogs took this all in stride as just another fun activity they did with their owners, and their brains responded much as a human's would to a pleasurable experience.

As the dogs became increasingly accustomed to the MRI, we were able to design more complex tasks for them to do. When we presented smells of people and other animals to the dogs, we found that the reward response occurred in the dogs' brains only for the smells of the people in the household, and not for the smells of the other dogs. Because food wasn't directly linked to these odors, this was the first hard evidence that dogs might experience something like love for the people in their lives.

The Dog Project soon consumed my life, eclipsing the human work in the lab. Because of its potential to improve the training of military working dogs, the Office of Naval Research began supporting our research, and we expanded the number of dogs in the project as well as the complexity of the tasks they performed in the MRI. Not only was it fun, but I had the feeling that we were on the verge of gaining new insights into the minds of our best friends.

As I learned more about the canine brain, I became convinced we had much in common with dogs at the deepest levels. The same basic structures for emotion could be found in both dog and human. But there was a bigger question here, beyond one of emotions, which I had conveniently suppressed as we had built up the Dog Project.

The question came to the surface at a conference on vegan issues. I had been reluctant to accept the invitation to speak because I wasn't a vegan, but the organizers assured me that they just wanted to hear what we were learning about the canine mind. Personal eating practices were not on the agenda. That might have been the plan, but it didn't turn out that way. After my talk on the Dog Project, a fellow speaker accused me of being a "speciesist" because I gave dogs special status, even to the point of feeding them the ground-up flesh of other animals in the form of hot dogs. It was an awkward moment, and I felt duped by the conference organizers.

Was I a speciesist? Probably.

Was that bad? I didn't know.

Four years into the project, there was no denying that our work had raised a bigger question: If we had evidence that dogs experienced emotions similar to those of humans, what about other animals?

People began asking me whether cats could be trained to go into the MRI, and occasionally about whether pigs might be trainable. I knew that wasn't likely, and scanning sedated animals didn't seem ethical or likely to yield much useful information about animal cognition. I was at an impasse: the possibility of studying other animals seemed a fantasy.

The turning point came when Peter Cook joined the lab. Peter had come to the project from Santa Cruz, California, where he had completed his PhD on sea lion memory. He was passionate about figuring out how animals' minds worked, especially in their natural environments. California sea lions, though, had been stranding in large numbers. Some of the sea lions could be rehabilitated, but others suffered from unremitting seizures and had to be euthanized. Peter arranged for us to get their brains. I never imagined I would be in the business of scanning dead brains, but I was surprised by what we learned about the animals they came from. It was some comfort to know that even in death, these animals could tell us something about the worlds they had inhabited. The sea lions were just the beginning. Using new MRI techniques, we started pushing the boundaries of what we could scan. Other animals. Specimens locked away on museum shelves. And even the brains of animals that were thought to be extinct.

What is it in a human brain that makes a human, or in a dog brain that makes a dog? For centuries, anatomists focused on size. Bigger meant more neural real estate, and the assumption had been that bigger was better. This principle had been taken to apply to the whole brain, where bigger brains were associated with greater intelligence. And it had also been applied to parts of the brain, where the size of specific structures was thought to indicate the importance of the function of that region to the animal. There was some truth to this. Dogs had a large olfactory bulb, indicating the importance of smell in the dog's world.

But size alone does not explain how brains work. What really matters is how the different parts of the brain are connected to each other. This is the new science of *connectomics*. Recent advances in MRI have allowed us to examine in exquisite detail the wiring diagram of human brains. If I, or anyone else, were to ever figure out animals' minds, it was going to come from the analysis of these connections and how they coordinated activity throughout the brain. That was where internal experiences, including emotions, came from.

These were exciting times to be a neuroscientist, and the Dog Project was just the beginning. The deeper I got into the dog brain, the more obsessed I became with learning about other animals. If we could learn about their inner experiences, maybe we could communicate with them better. What if a dog could tell us exactly how she felt? And what would a pig say about a slaughterhouse? What did a whale think about all the noise flooding the ocean from ships and submarines? The inevitable result of these investigations was not just that we were going to realize that the inner worlds of animals are a lot richer than we had imagined, but also that we need to rethink how we treat them.

This is a book about brains, and about the minds of the animals they come from. In academics, such investigations fall under the umbrella of *comparative neurobiology*. All neuroscience is comparative at some level, but few neuroscientists dig deep and ask why the brains of animals look the way they do and how that relates to their mental experiences. These are hard questions. They get at the heart of what makes us human, and they raise troubling issues about the possibility that we may not be that different from many of the creatures with whom we share the planet.

The book is organized in roughly the sequence in which I branched out from humans to dogs to other animals, but the similarity of brains is the thread binding these adventures together. Over and over, I found structures in the brains of animals that looked to be organized in the same way as the corresponding parts of our own brains. And not only did these parts look the same, but they functioned in the same way.

The relationship between brain structure and cognitive function is complex and frequently depends on the coordination of multiple brain regions. Until recently, it wasn't possible to describe in detail the interconnectedness of the brain. But this has changed in just the past few years. Advances in neuroimaging and the software used to analyze brain networks have yielded new insights into human brain function, and there is no reason why the same tools can't be applied to animals' brains.

These techniques also suggest a way to understand the subjective experiences of other animals. Where structure-function relationships in an animal's brain are similar to those in our brains, it is likely that the animal is capable of having a similar subjective experience as we do. This, I believe, is the path toward understanding what it's like to be a dog, or a cat, or potentially any animal.

Dogs figure prominently in several chapters because dogs are familiar to all readers and because I think they are the best research partners. I also venture into the ocean to discover what the minds of our marine relatives are like. There are chapters on the most doglike of marine mammals, sea lions and seals, and a chapter on one of the most mysterious animals on the planet: dolphins. With their great intelligence and sociality, dolphins have intrigued both the public and scientists for decades. But they have long been inscrutable. Now, using new imaging techniques, we are learning how dolphins' brains are wired and what this means for life underwater. We may soon be able to communicate with each other.

And then there is the Tasmanian tiger, officially known as the thylacine. The "Tassie tiger" was a carnivorous marsupial that looked strikingly like a small wolf. It is believed to have become extinct in 1936, when the last one died in the Hobart Zoo in Australia. But claimed sightings of this mysterious creature continue to this day. I began a quest to find intact thylacine brains to see what I could discover about their inner lives, and ultimately, I located a preserved brain in the vaults of the Smithsonian Institution—one of only four known to exist in the world. I received permission to scan it with the new MRI tools. But that was just the beginning of an odyssey that took me to Australia in search of more brains and to scan the thylacine's closest living relative, the Tasmanian devil.

The book ends as it begins—with dogs. Although I admit to being an unabashed speciesist, I have come to think of dogs not just as man's best friend, but as ambassadors to the animal world. They still have enough wolf in them for their brains to tell us something about being a wild animal. The challenge lies in creating a means to communicate with each other. I believe that we should look to their brains. And so, the final chapters are about probing the limits of canine understanding of human language and what that means for the rights not only of dogs, but of all animals.

Chapter 1

What It's Like to Be a Dog

In the early spring of 2014, a dedicated group of volunteers coaxed their dogs into a simulator of an MRI scanner.

While waiting his turn, a big, yellow dog named Zen bounded over to me and lowered his head while raising his butt in the air. He wagged his tail, insisting that I play with him. I obliged Zen's request and tussled with him. After a few minutes of gentle tumbling, Zen had had enough and demonstrated why his name was so apt. He sat his rump on the floor and, in no particular hurry, let his front paws slide out in front of him. He stared back at me, looking as peaceful and inscrutable as the Sphinx.

I wondered what it was like to be Zen.

Zen was a cross between a yellow Lab and a golden retriever, and he was one of the veterans of the Dog Project. As a puppy, he had been slated to become a service dog, but when he entered adolescence, his handlers thought he was too distractible for assistance duty. He was released from the program to be adopted by his puppy-raiser. By tradition, all the dogs of a

9

Zen. (*Gregory Berns*)

litter have names beginning with the same letter. Zen just happened to be born into a litter assigned the letter Z. Whoever named him could not have had any idea of his future personality. Maybe dogs grow into their names, but the fact that his name captured his personality seemed like some kind of karmic coincidence.

A diverse group of dogs and humans filled the practice space. Zen and his fellow service-dog washouts formed a loose grouping on one side of the room. His buddies included Pearl, a compact and energetic golden retriever who had also been released for distractibility. There was Eddie, short for Edmond, another Lab-golden cross, who could have been Zen's twin but was released for a predisposition to hip dysplasia. Ohana, a pure golden retriever, was slightly less kinetic than Pearl. Kady, a sweet retriever mix, had been released for being too shy. And then there was Big Jack, a phlegmy, one-hundred-pound golden retriever getting on in years, who was happiest when getting a steady supply of hot dogs.

On the other side of the room, Peter Cook, a postdoctoral fellow from Santa Cruz who had studied sea lions, was supervising a second group of dogs who were less humanized than the retrievers. This crew of irrepressible dogs was led by Libby, a liver-brown pit mix with a limp tail who was holding as still as a statue in the chin rest we'd designed to help the dogs keep their heads in position during imaging. Her owner, Claire Pearce, had found Libby wandering on the side of a highway in California. It was only because Claire was an experienced animal trainer that Libby had become socialized enough to be around people. But being around other dogs was another story. They still sent Libby into a fit of barking and lunging. Claire had staked out a corner of the room where she could control Libby and where Libby wouldn't disturb the other dogs.

Although many of the other humans participating in the Dog Project didn't like Libby, I was fond of her. She reminded me of Callie, the black terrier mix whom my wife had adopted from the humane society. Even though Callie was emotionally distant, insecure, and had a tendency toward bullying, she was eager to work. She'd been the first dog to train for the MRI, and the bond we had formed during the project was as intense as any I had ever had with a dog.

Zen and his gang of retrievers were great dogs, the kind every child would love to have, but dogs like Libby and Callie were slightly less domesticated, slightly wilder creatures. They seemed like throwbacks to the last ice age, when our prehistoric

Libby. (*Gregory Berns*)

ancestors bonded with wolves and turned them into dogs. Living with dogs like Libby and Callie meant accepting unpredictability. Whether these differences in personality came from genetics, levels of puppy socialization, or variations in brain function, nobody could say, but I aimed to find out what it was in Zen's brain that made him Zen, and what made him different from Libby and the other dogs.

The undertaking was not without controversy. Many academics rejected the idea that we could know the mind of an animal, even with modern neuroscience techniques. The crux of the problem was an influential essay by the philosopher Thomas Nagel titled "What Is It Like to Be a Bat?" Neuroscience, Nagel had said, could never explain the subjective experience of having thoughts and feelings. Even if we knew how a bat's brain worked, it wouldn't get us any closer to what it would be like to be a bat. Bats were just too different from humans. Consider sonar. Because humans had no such faculty, we could never imagine what it was like to be a bat using sonar. And forget about flying. According to Nagel, nothing in the bat's brain could tell us what it was like to fly.

Nagel's essay cast a long shadow over the interpretation of neuroscience data. Neuroscience dealt with the measurable properties of the brain, but subjective experiences were not so easily quantified. No instrument existed to measure the full experience of smelling a rose or what a dog felt when his owner came home. And the harder we tried to pin down the objective qualities of these experiences, the further we moved away from the unique subjective experience of what they are like. Without the means to quantify the subjective experience, there could be no marriage with neuroscience. According to Nagel, we could deconstruct brains all we wanted, but without the link between subjective and objective, we would never get any closer to knowing what it was like to be an animal. The argument applied to humans, too. No matter what we said or did, there would be no way to fully know what it was like to be someone else without actually being that person. According to this logic, looking in a person's brain wouldn't help.

His two examples, flying and echolocation, at first glance do seem very different from human experience. But thrill seekers now regularly don wingsuits and sail through alpine canyons. They look like human bats, and those who dare to take flight can tell us what it is like to fly. Even the echolocation argument falls flat. We all have a nascent ability to sound out a room.

Simply by speaking, it is not difficult to make out the differences in size and composition of a bathroom, a dance hall, or a concert auditorium.

When we ask what it's like to be a bat, or a dog, we are asking about the internal experience of an animal. Call it a mental state. The question is one of internal versus external perspectives. Nagel argued that we couldn't know what it's like to be a bat (or another person) without being that individual, because subjective experience is an internal perspective—how an individual feels on the inside—and that is different from how they might describe the feeling to someone else or what another person might observe about them. The description of internal feelings is one way to share experiences with each other, but as Nagel pointed out, they are not the same thing as the experience itself.

But just because we can't be another human being doesn't mean we can't have a pretty good idea of what it's like to be someone else. Language plays an important role, allowing us to communicate and describe stuff to each other, but even language may not be all that necessary for the sharing of experiences. The main reason we can describe events to each other is that humans share the same physical attributes and inhabit the same environments. We're so similar that language can ride on top of these commonalities, acting as a symbolic shorthand.

These commonalities extend to other animals. We share basic physiological processes necessary for life with many different types of animals, and within the class of mammals, we share even more. We all breathe air. We have four limbs. We sleep. We eat. We reproduce sexually and give birth to live young that nurse for a period of time. And many mammals are highly social. With such physical similarities, the structure of our internal experiences is not likely to be as different as has often been assumed.

These physical domains suggest a way to understand the internal experience of another individual. Instead of trying to answer the big question of what it is like to be a dog, we can be more precise. What is it like for a dog to experience joy? Even more specifically, what is it like for Zen to experience joy? Or what is it like for Libby to refrain from barking at other dogs? Obvious domains in which we could ask questions include perception, emotion, and movement. There are also domains necessary for the maintenance of bodily functions, such as sleep, thirst, and hunger. The sum total of all of these domains constitutes mental experience.

Humans have a few extra dimensions, notably language and symbolic representation. Apart from communicating with each other, language lets us conduct an internal monologue. It rides on top of the other domains, labeling other aspects of experience. We can't help it. Some have argued that language is so integral to human experience that words change everything. William James, the father of American psychology, wrote that a man is afraid of a bear only because he becomes aware that his heart is beating faster and says to himself, "I'm scared!"

The primacy of language had caused many researchers to abandon the possibility of knowing what an animal experienced. Because a dog can't say to himself, "I'm scared," some scientists had even taken to redefining the most studied animal emotion—fear—as a behavioral program that an animal implemented to avoid something painful. This was a step backward toward a Cartesian view of animals as automatons.

Some might think that a scientist should remain agnostic, but the same wait-and-see attitude had dominated the debate about climate change. It is true that much remains unknown about the climate, but at a certain point the evidence becomes too much to ignore, and any rational person will come to the conclusion that the planet is heating up because of human activity. The same is true for the mental lives of animals. As with climate change, there were consequences for denying their existence. Continued agnosticism about animal emotions, or even the degree of consciousness in different animals, had allowed people to exploit them in myriad ways. But this was beginning to change.

Before modern neuroscience techniques became available, the only way to gain access to mental states was by observing behavior, or, in the case of humans, by asking what a person was thinking or experiencing. Both are imperfect measures of mental states. The observation of behavior requires us to make assumptions about what an individual is experiencing internally. This works pretty well with people because of our physical similarities and shared culture, but with animal behavior we have a larger gulf to bridge to their internal states. And what if an animal isn't doing anything? How can we know what he or she is feeling, if anything at all? These types of questions were at the heart of Nagel's argument against the possibility of knowing what it was like to be an animal.

There might have been self-serving reasons for scientists to conclude that animals didn't experience emotions. Scientists had to justify invasive research procedures. But I found such rationalizations self-serving and disingenuous. The inability of animals to label an internal state did not mean they didn't experience something on the inside akin to what a human would experience under similar circumstances. And I wasn't the only one to question the status quo. With the benefit of forty years of progress since Nagel's essay, the pendulum was now swinging in favor of neuroscience. Two recent advances had shown that we could, in fact, use the brain to know what it was like to have mental experiences even in the absence of outward behavior.

In 2006, Adrian Owen, a Cambridge neuroscientist, used functional magnetic resonance imaging (fMRI) to study the brain responses of a twenty-three-year-old woman who was unresponsive after a traffic accident had caused severe brain trauma. By all clinical measures, she was in a vegetative state. Yet, when Owen and his team spoke to her, they observed increased activity in the left frontal cortex, especially to sentences that had ambiguous meanings. Even more remarkably, when the woman was given instructions to imagine herself playing tennis or visiting the rooms of her house, Owen had observed increased activity in the regions of her cortex associated with spatial navigation. Owen's results were hugely important. They demonstrated that internal subjective experiences could become disconnected from outward behavior, but that brain imaging could reveal the inner side.

In 2008, Jack Gallant, a psychologist at the University of California, Berkeley, pushed the boundaries of brain-decoding even further. Gallant demonstrated that he could determine what a person was looking at just by measuring activity in the visual cortex. Over the next several years, Gallant increasingly refined the technique to the point where he could determine not only what a person was looking at, but the type of image it was (such as a person, object, or scene), and even when a person was remembering an image instead of seeing it. Gallant's techniques proved that, without a doubt, measures of physical activity in the brain could be translated into discrete mental states—in this case, visual imagery. It was a triumphant vindication for the material reductionists. Specific mental domains could be decoded from the brain.

If these techniques worked for humans, there was no reason why similar approaches couldn't be used to decode animal states of mind. Knowing what it was like to be a bat, or a dog, began to seem like a real possibility.

With Libby warmed up, Claire released her from the chin rest. Libby saw me watching her and took that as an invitation to play. She tried to accelerate to full speed, but without traction on the slippery floor, she half-ran, half-hopped in excitement over to me. She leapt toward my face. Anticipating this, I dodged to the side, and Libby went whizzing by. Only then did I kneel down to accept her aggressive face-licking.

Claire came over and clipped on Libby's leash. "Libby. Enough."

Libby sat her butt on the floor and swiveled her head back and forth between Claire and me. It took every ounce of self-control for Libby not to leap at my face in slobbery excitement. If she didn't have a limp tail, it would have been sweeping the floor.

"Let's put her in the tube," Peter said.

The tube was a six-foot piece of Sonotube, which normally would have been used as a mold for concrete pillars. I had re-purposed it to simulate the inside of an MRI scanner. It was mounted horizontally on a table in the center of the practice space and had a piece of plywood inside of it to serve as the patient table.

Claire led Libby to a set of portable steps that ascended to the opening of the tube. Libby had been participating in the Dog Project for three years, and she knew what to do. She rocketed up the steps and scrambled into her chin rest, a block of foam with a semicircular cutout to match the profile of Libby's muzzle. The chin rest was affixed to a mockup of the head coil, the part of an MRI that picked up signals emanating from the brain. The human version looked like a Stormtrooper helmet out of *Star Wars*, but for the dogs we only used the lower part of the coil, the piece that normally sat around a person's neck.

With Libby in a crouch position and her head in the coil, Claire began to run through trials of a new experiment. All of our previous studies had been passive tests. In those experiments, we had presented stimuli in the form of hand signals, computer images, treats, and smells. The dogs didn't have to do anything other than hold still while we measured their brain responses, and the experiments had been tremendously successful. We had trained more than twenty dogs in this way, and we had published several papers on how the reward center of the dog's brain worked. But now Claire and Libby were practicing something far more complex. They were working on an active task. Libby would, for the first time, perform a behavior during a live MRI scan.

Studying dog behavior while scanning ran counter to all of the requirements for motionlessness that we had established in the passive studies, but it was the key to understanding what made Libby *Libby*, and what in her brain made her different from Zen. Their outward behaviors made clear that they were different from each other, but we couldn't very well have them in the MRI reacting to dogs and people. Maybe Zen would have remained still with characteristic aplomb, but Libby sure as hell wouldn't have.

So we borrowed an experiment from the field of human psychology, something that even children could do. It was called the Go-NoGo task.

With Libby in the simulator, Claire took out a plastic dog whistle.

Claire stared down the tube at Libby and blew.

Without hesitation, Libby nudged a small plastic target taped to her chin rest about a centimeter from her nose. Claire pressed a button on a palm-sized box, making a loud click. The clicker indicated to Libby that she had done the right behavior, and Claire rewarded her with a treat.

So far, so good. Libby had learned that the whistle meant: *Poke the target.* For most of the dogs, this had been remarkably easy to train. We had started with targets on the floor. Just by pointing to the target, we could signal to the dogs that they could investigate, which they would do happily. It was a simple matter to blow the whistle at the same time as pointing. When a dog touched the target, the owner rewarded the dog with a treat. Pretty soon, the pointing became unnecessary.

Next came the hard part. With the whistle still in her mouth, Claire raised her arms and crossed them in the shape of an X. This meant: *Don't move. Even when you hear the whistle.*

With her arms crossed, Claire blew the whistle at a low volume.

Libby stared impassively.

"Good," Peter said. "Reward her."

To make sure this wasn't a mistake, Claire lowered her arms and blew the whistle again. Now, Libby poked the target.

"Good girl!" Claire exclaimed while giving her another treat.

It appeared that Libby understood that the whistle meant *Go* and crossed arms meant *No-Go*, overriding the whistle.

"That looks great. Let's go up on the whistle volume," Peter said.

Repeating the drill, with arms crossed and the whistle louder still, Libby continued to hold still in the chin rest. I was thrilled. This is a difficult task even for humans.

The Go-NoGo task has been a staple of the psychologist's toolbox for decades. We were using the whistle to signal Libby to perform a nose-poke, but a human doing this task has to press a button on a keyboard. Even humans require a fair bit of self-control to do a Go-NoGo task, and individuals vary in their ability to do it well. Young children, lacking development of the frontal lobes, can't do it at all. I hoped this individual variability would also be present in dogs, because that would open a window into the differences in their brains.

Dogs don't have very big brains. They are about the size of a lemon, and relatively speaking, the frontal lobes are far smaller than those in humans. It is no surprise that dogs aren't great masters of self-control. Sure, dogs can learn tricks and even sit for long periods of time waiting for their humans to toss them treats, but around my house, our dogs are constantly on the prowl to swipe forbidden items like food and underwear. Even though she could barely reach the top of the kitchen counter, Callie was skilled at lapping up morsels of food by tipping her head sideways and extending her tongue like an anteater. Either she couldn't help herself, or she had immense self-control and knew just how far the humans could be pushed before someone yelled at her.

Self-control, or rather the lack of it, often lands dogs in shelters. Biting, barking, property destruction, and household urination are the most common reasons people give up their dogs. Understanding which parts of the dog brain are responsible for impulse control and how these regions work had become a major goal of the Dog Project. If we could make progress in this area, we might be able to decrease the number of dogs surrendered to shelters that get euthanized.

Libby was confounding the team. She had little impulse control around other dogs, but became a model canine citizen while in the MRI simulator. Our simple idea of self-control as something a dog either had or did not have was incomplete. If Libby could have great self-control in one situation but not another, it had to be dependent on context. We wanted to know how.

Although the other dogs weren't as excitable as Libby, many of them had a difficult time learning the Go-NoGo task. It took some of them several months to achieve Libby's level of proficiency. I couldn't fault the dogs. After all, many had been participating in the MRI project since its inception. When they were in their head coils, whether in practice or in the real scanner, they were not to move. The Go-NoGo task went against all their training. And although Libby seemed to adapt to the new circumstances, the other, more passive dogs seemed stuck in their old habits. I didn't know if they were stubborn or confused.

Kady typified this mental inertia. Like Zen, she was a cross between a golden retriever and a Labrador. She had a dense blonde coat, almost snow-white, which set off her big chocolate-colored eyes. Kady was one of the sweetest dogs I had ever met, but I also had to admit that she seemed rather vacuous, adrift without purpose unless her owner, Patricia King, was there to tell her what to do. No doubt much of this was genetic. As a potential service dog, Kady had come from a long line of dogs selected to do the bidding of humans who could not do things themselves. It was almost as if Patricia were Kady's auxiliary brain. For dogs like Kady, it wasn't clear whether we could separate her wishes from her owner's, or whether that was even a meaningful distinction.

Behavior alone wouldn't tell us what it was like to be Kady, because behavior could be driven by a variety of motivations. Only by reading out the state of her brain could we gain insight into why she did something or not. And with Libby and Kady at opposite ends of the spectrum of dog compliance, we had a perfect opportunity to sort out the interplay between what a dog wanted and what was allowed.

But first we had to get Kady to engage in this new type of task.

Kady didn't even want to touch the target. This was strange, because Kady's favorite game was to play ball. I had thought that chasing a ball would be similar to nose-poking a plastic target. Both activities involved doing something with the nose and mouth. But, once again, I had fallen into the trap of thinking like a human instead of a dog.

When Kady was in her head coil, she never moved, which was why she had been our best and most consistent dog in the project. It had made scanning a breeze. The downside was that as soon as we changed the context of the task, Kady shut down. It appeared that rather than being able to figure out what was required of her, she had opted to go immobile and wait for cues from Patricia.

Fortunately, we could piggyback on dogs' preternatural ability to follow human pointing. Brian Hare, an evolutionary anthropologist at Duke University, had studied this ability in several species, including dogs and primates. He found that dogs seemed to know that when a human pointed at something, they should look at the thing being pointed to. To a human, this may seem obvious, but other primates could not do it without a lot of training, assuming they could learn it at all. A monkey might just stare at your finger. Researchers debate whether following a point is innate or learned. Monique Udell, a canid behaviorist at Oregon State University, showed that wolves raised by humans since birth did just as well as dogs on pointing tasks. She argued that human hands carry special significance to dogs and socialized wolves, because the canids quickly learn that human hands often dispense treats. In her view, it is a natural extension of the significance of the hand to look where the hand points, especially when the hand itself doesn't contain food. Hare, conversely, argues that the behavior is innate, and that it has been bred into dogs over millennia. Regardless of whether the ability to follow a pointed finger is innate or learned in dogs, however, we used it to teach Kady what to do.

First, with the plastic target on the floor, Patricia pointed to it. Kady went to Patricia's finger and nosed around with her butt in the air and tail wagging. No doubt she expected a treat. In the process of nosing around, Kady inadvertently knocked over the target. This of course was the goal, and as soon as she did, Patricia said, "Good girl!" and gave her a treat.

Kady didn't yet know what had happened, but she knew it was fun.

And so they repeated this exercise again and again.

After about twenty repetitions, it finally clicked. When Patricia pointed to the target, Kady ran over to it and knocked it down with purpose. Only after hitting the target would she look at Patricia for her praise and treat. She had learned that treats were contingent on knocking down the target.

Once Kady learned that knocking down a plastic target was a fun game, Patricia introduced the whistle. Instead of simply pointing at the target, Patricia blew the whistle at the same time as pointing. Once again, after another twenty repetitions, Kady transferred the association to the whistle, and Patricia didn't need to point at all.

With Kady reliably nose-poking, it was time to transfer the target into the head coil. For this experiment to work, all the dogs, timid or not, would need to nose-poke the target in the head coil in the MRI. But dogs are surprisingly sensitive to context. Just because Kady had learned to nose-poke on the floor didn't mean she would do the same thing in the head coil. Even though it seemed obvious to us that the target was the same, we had no guarantee that Kady would see it that way. We didn't yet know what it was like to be Kady.

With Kady in her head coil and the target taped to her chin rest about a centimeter from her nose, Patricia blew the whistle. And Kady just stared right back at Patricia. I couldn't read any expression in Kady's eyes.

Patricia tried a few more times and looked up in exasperation. "What do we do now?"

The answer, of course, was to do exactly what she had done on the floor. Point and, if necessary, tap the target in conjunction with blowing the whistle until Kady understood what we wanted her to do.

For the dogs like Kady it was slow going. Unlike Libby, they just didn't want to move in their head coils. In the end, all the dogs were trained to nose-poke on the whistle and, to varying degrees, inhibit the nose-poke when the crossed-hands signal was raised. Some dogs got it down in a couple of months, while others took six months.

Then it was on to the scanner. If we were lucky and had properly designed this experiment, we would soon find out what was going on in the minds of these dogs.

Chapter 2

The Marshmallow Test

Patricia liked to go to the MRI early, so she and Kady got the first slot on a beautiful day in April 2014. I had arrived thirty minutes earlier to boot up the scanner computers and get the room ready for dogs. I put fresh sheets on the patient table and placed the pet steps at the foot of the bed so the dogs could walk up into the bore on their own. Peter came a few minutes later and began setting up his equipment at the rear of the magnet. First was the button-box that Peter would use to time-stamp each trial. He would push one button when Patricia blew the whistle and another when Kady nosed the target. Next, Peter taped a mirror inside the bore of the magnet so he could see when Kady actually touched the target. Finally, we duct-taped Kady's chin rest in the head coil, and, with the push of a button on the side of the scanner, sent the whole thing into the center of the magnet.

Kady bounded into the control room, tail and butt wiggling from side to side. This was the tenth time she had done this. Not a trace of anxiety. After a few minutes to let her settle down, Patricia told Kady, "Coil!"

Kady bounded up the steps, plopped her head in her chin rest, and waited for Patricia to come around to the rear of the magnet where they could face each other. The plan was to warm up by running through ten Go and ten NoGo trials without actually scanning. We didn't expect perfect performance, but anything near 80 percent accuracy on both trial types would be good.

But Kady had reverted to her *I do not move in the MRI* mode. No amount of whistling could get her to budge. Time for problem-solving. Mark suggested that we take her out of the scanner and get her back into play mode.

"Release!" Patricia said.

Kady backed out and walked down the steps, looking as if nothing were amiss.

We put some plastic targets on the floor just like when we'd first taught her. Even though we liked the dogs to be calm, in this case I played into Kady's personality and started tussling with her to get her more excited and wanting to play. Patricia whistled and pointed to a floor target. Kady ran to her, knocking over the targets. We all cheered.

After ten minutes of bowling for targets, Kady seemed ready to try again in the scanner. This time it worked. Although she wasn't perfect, she nose-poked the target on about 75 percent of the whistles. We were ready to go live.

MRI is a wonderful tool. It is, by far, the best technology for looking inside the body. No X-rays or other forms of ionizing radiation are required, just an extremely powerful magnet and a lot of highly engineered software to construct images. And because no radiation is involved, MRI is safe.

But MRI scanners can be temperamental. Parts fail and mysterious errors appear on the control panel, requiring the attention of the secret order of technicians known as MRI repairmen. The main source of this complexity is the requirement that the magnet be kept cold. Everything depends on generating a magnetic field 60,000 times as powerful as the Earth's. And the only way to do this is by wrapping miles of electrical wire around a tube and then sending an electrical current down the line. The electrical current creates a magnetic field aligned with the bore of the tube. The problem is that it takes a lot of electrical current to generate big magnetic fields—so much current that it would melt any copper wire. It wasn't until the 1970s that this problem was solved by the discovery of a special class of materials known as superconductors. MRIs use wires made of niobium and titanium. When these exotic metals are cooled to extremely low temperatures, they lose all electrical resistance and can handle as much current as needed. Because they have no electrical resistance, the wires don't heat up and the current doesn't dissipate. Once it is energized, a superconducting magnet is always on.

These low temperatures can only be supplied by liquid helium. Normally, helium exists as a gas that is lighter than air. But if you cool it down enough, eventually it will condense out of its gas phase into a liquid form. This happens at around −269°C (−452°F). Left on its own, the helium would simply boil off into outer space. You have to keep a lid on the whole system, like you do with a pressure cooker. Even then, the laws of thermodynamics dictate that the helium will slowly revert to gas form. A pump keeps compressing it to maintain the liquid state as long as possible.

All of the pumps and tubes give the impression that the MRI is alive. The compressor, which is called the cold-head, sounds like a ventilator. When you walk into the magnet room, the first thing you notice is the kachunk-kachunking of the cold-head. It never stops. At least it isn't supposed to. When we installed our new MRI scanner in the psychology building at Emory University, it unexpectedly vented helium, which caused the loss of superconductivity and the collapse of the magnetic field, a costly event called quenching. Because of our magnet's volatile temperament, we named the scanner Penny, after the similarly unpredictable character on *The Big Bang Theory*.

A lthough the magnetic field of the MRI is always on, the actual scanning involves the addition of small amounts of magnetism in precise locations. The additional fields are controlled by auxiliary magnets called gradient coils and are located just inside the bore of the scanner. When you send electrical current through the gradient coils, you can isolate specific locations in the brain. Sophisticated software controls the gradients so that each location in the brain is isolated in rapid sequence.

If the gradients operated behind the scenes, none of this would matter. But the rapid cycling of the gradient fields is loud. Every time the gradient currents are changed, the coils vibrate, and these vibrations get transmitted through the entire scanner. The MRI acts as a massive loudspeaker with our dog participant smack in the center of it.

Early on in their training, we acclimated the dogs to the scanner noise by playing recordings of it at the appropriate volume. But the gradients are so loud that the dogs still needed hearing protection. We used the same type of foam ear plugs for the dogs as we gave to humans going into the scanner, except the dogs' ear plugs were held in place with a colorful wrap. And

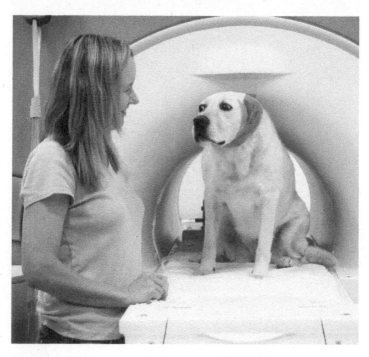

Patricia wraps Kady's ears in preparation for scanning. (*Helen Berns*)

for the dogs who didn't like things in their ears, we gave them ear muffs.

With our participant wrapped and Peter in position, Patricia commanded, "Kady, coil!"

From the control room, all I could see was Kady's rear. When she looked comfy, I checked her positioning with a localizer scan. This is a ten-second scan that gives a snapshot of whatever is in the coil. For this type of scan, the gradients make a low-pitched buzzing that doesn't seem to bother many of the dogs. As expected, Kady's brain was dead center in the field of view.

Next up were the functional scans. This was the sequence

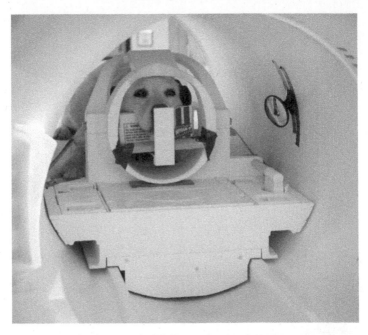

Kady settles into her chin rest in the MRI. (*Gregory Berns*)

that would capture Kady's brain in action. With functional MRI, or fMRI, the scanner is programmed to acquire brain images rapidly and continuously. How rapid depends on how big the brain is. For humans, it takes about two seconds to acquire the whole brain, but because dog brains are only the size of a lemon, we can capture the whole thing in half that time. The end result is a sequence of images not unlike a movie. Because fMRI cycles the gradients at high speed, the sound pressure is ninety-five decibels, which is equivalent to the noise of a jackhammer at a distance of fifty feet. Hence, the necessity of ear protection.

I hit the scan button on the control panel. The

jackhammering cued Kady and Peter that we were live.

As Patricia put up the first hand signal, images of Kady's brain began streaming to the console. As functional images, these did not contain a lot of detail. The scans were optimized to pick up changes in the oxygen level in the blood vessels surrounding the neurons. When neurons fire, the neighboring blood vessels dilate, letting in fresh blood for the neurons to recharge their energy stores. In fMRI, the scanner picks up the changes in blood flow, revealing the location of neural activity. It is called the *blood oxygenation level dependent* response, or BOLD for short.

The BOLD response is tiny—less than 1 percent of the total signal. To make matters worse, the fMRI signal itself is noisy, bouncing around by 5 to 10 percent. Some of this noise comes from the thermal motion of water molecules, but most of it is physiologic in origin. Pulsing blood causes the brain to move with every heartbeat. The respiratory cycle induces motion in the head while altering the concentration of oxygen and carbon dioxide in the blood. These sources of noise conspire to swamp the BOLD signal. Fortunately, the law of large numbers provides a way to overcome noise by averaging many repetitions. Random noise drops by the square root of the number of repetitions. So 100 trials in the scanner would drop the background noise by a factor of 10.

Although it was hard to see over Kady's body in the MRI, I could hear the dog whistle through the intercom. If Patricia didn't give the NoGo signal, Kady usually poked the target, which was immediately evident on the stream of MRI images by a sudden shift forward of her head. I watched closely to make sure that after each nose-poke, Kady's head returned to the same position.

After ten minutes of scanning, Peter came out from behind the machine and waved. They were finished with the first set of trials, and I stopped the scan. Kady backed out of the bore and trotted down the steps, wagging all the way.

Peter asked, "How did she look?"

"Good," I replied. "No head movement between trials."

"That's because she was being conservative."

Conservative meant that Kady had a tendency not to move when she was supposed to. That was just her tentative personality. Kady had a self-initiative problem. After a five-minute break, it was back into the scanner. We did the cycle three more times, and then Patricia and Kady were done. In the end, Kady had a hit rate of 75 percent on the Go signals, but she also had a surprisingly high false alarm rate of 56 percent on the NoGo signals. This meant she nose-poked about half the time when she wasn't supposed to. In addition to her lack of self-initiative, she also exhibited self-control problems. We hoped her brain would tell us why.

By that time, Claire and Libby had arrived. Libby zoomed around the control room, leaping up at everyone. Based on Libby's personality, I expected her to have even less self-control than Kady. Once again, my intuitions were wrong.

If anything, Libby was even more conservative than Kady. Libby had the same tendency not to move when in the scanner. We proceeded through the same warm-up drill on the floor that we did with Kady to get Libby in the mode of nose-poking targets, then told her to enter the scanner. Libby's scores were mixed. While she had been almost perfect on both Go and NoGo trials in practice, in the scanner she only poked 46 percent of the Go trials. But her accuracy on the NoGo trials was a near perfect

96 percent. So whereas Kady's performance seemed to deteriorate in the scanner on both types of trials, Libby just became more conservative, which only affected her performance on Go trials. Her performance on NoGo trials improved.

Big Jack was one of the dogs who performed well on both trial types. At nine years old, he was the oldest dog in the project. Apart from being seriously overweight, his age had begun to slow him down. It was always touch-and-go to get him up the steps into the scanner. I stood on one side of the patient table to stabilize him if he started to slip off. The good thing about Jack, though, was that once he was on the table, he stayed there for the entire session.

Jack was nearly perfect in his warm-up, so we went straight to scanning. He and his owner, Cindy Keen, breezed through the experiment. He was so good at this that he only needed one break. He had a 70 percent hit rate on the Go trials and a remarkable 96 percent on the NoGo trials.

Over the course of several months, we scanned thirteen dogs on the Go-NoGo task. Considering that the Dog Project had begun with my dog Callie and one other dog just a few years earlier, I was proud of how far we had come. The number of dogs trained for the MRI continued to grow, and the complexity of the tasks they were performing was approaching that of human fMRI studies. Although Callie had continued to participate in all the experiments, including the Go-NoGo task, I much preferred operating the scanner than being an active participant with Callie in the scanner. This was due in part to the increasing demands of directing the project, which ate into the time I had to train with Callie. But because Callie and I had been the first dog-human team to accomplish canine fMRI with the dog fully awake, I also felt pressure to set the performance standard whenever we were in the scanner.

Callie seemed to do well with the Go-NoGo task. She had a hit rate of 83 percent on the Go trials and was successful at inhibiting her response on 89 percent of the NoGo trials. Only Eddie did better. But once we looked at her scans, we found that Callie was moving too much during the waiting periods to produce useful data.

Movement wreaks havoc with the interpretation of fMRI findings. If the dog's head moves while a scan is being acquired, then the signal in adjacent brain regions gets mixed together. How much movement is tolerable depends on the resolution of the scans. Most fMRI experiments scan the brain at a resolution of 3 millimeters, which means that the brain is digitally diced into cubes measuring 3 mm per side. These are called voxels (short for volume elements), which are the three-dimensional analog to pixels (short for picture elements). Image artifacts appear when the magnitude of movement gets close to the size of the voxels. To be safe, we discarded scans in which the brain had moved by more than 1 mm from the previous scan.

Peter delivered the bad news about Callie. "After motion processing," he said, "she had less than a third left."

That was not enough to analyze. We would have to leave her out of that part of the experiment. I was disappointed, but I couldn't let my heart get in the way of rigorous analysis. And Callie wasn't alone. Ohana also moved too much.

With the remaining eleven dogs, Peter calculated the average BOLD response during the successful NoGo trials. Because the dogs were inhibiting the tendency to nose-poke, their heads weren't moving, and we were able to capture the brain activity responsible for putting on the brakes. For comparison, we had also included trials in which the owner held up one hand. This was a signal that all the dogs had learned from the very beginning of their participation in MRI training. One raised hand meant: *Hold still and you will get a treat.* This formed an ideal contrast to the NoGo signal, crossed hands, which meant: *Hold still, even when you hear the whistle, and you will get a treat.* Both conditions used hand signals, and in both cases the dogs received a treat upon success. The only difference was the addition of self-control when the whistle was blown.

By taking the average response during the successful NoGo trials and comparing it to the average response during the control trials, Peter isolated the parts of the brain that came online during active inhibition.

Only one region showed up: a small area of the frontal lobe.

Unlike the heads of humans, most of a dog's head is muscle, bone, and air. The muscles attached to the jaw and neck give dogs their powerful bites, while spaces in their skulls form cavernous sinuses for scenting. The brain, well protected beneath these layers, occupies maybe a quarter of the volume of the head. The frontal lobe is a small swath of brain behind the eyeballs. In humans, the frontal lobe takes up the front one-third of the brain, which is a lot, even for a primate. In dogs, it takes up only about a tenth of the brain.

In humans, the frontal lobe does lots of things: language, abstract thought, planning, and social intelligence as well as many other cognitive processes that we do not yet understand. But just because dogs don't have a big frontal lobe doesn't mean that they don't have analogous cognitive processes (although their lack of language seems obvious). The frontal lobe is defined by the anatomical boundary from the front of the brain rearward to a prominent groove. In primates, this groove runs from the crown of the brain toward a region adjacent to the ear. Anatomists call this groove the central sulcus. On the forward bank of this groove are neurons that control movement, while the rearward bank contains neurons that sense touch. The central sulcus is the dividing line between sensation and action.

Dogs don't have a central sulcus. Instead, they have a prominent groove from the top of the brain that runs forward in an inverted chevron, called the cruciate sulcus. The area that Peter identified lies at the lower part of this groove. Other researchers had identified activity in an analogous region in both humans and other primates during similar tasks, suggesting that we had, in fact, located the part of the dog's brain that, like its primate counterpart, inhibits the impulse to move.

But knowing the location of this region in the dog brain only served to confirm that we were on the right track. What we really needed to understand was how this region enables self-control. If we knew that, then we could begin to understand the differences between dogs like Kady, Libby, Jack, and Callie. Maybe we could even help dogs with self-control problems so that they wouldn't end up in shelters.

The first clue came from the dogs' varying levels of skill on the Go-NoGo task. Just like some humans, some dogs did better than others, and there seemed to be a direct correlation between the level of activity in the dogs' prefrontal regions and their performance on the task. Higher prefrontal activity was associated with a lower false alarm rate. This relationship suggested that the dogs who brought more of their limited frontal lobes to bear on the task did better than the ones who brought less. Our next task was to see if dogs who performed well on the Go-NoGo task would also do well on other tasks of self-control.

The most famous experiments of human self-control were led by the Stanford psychologist Walter Mischel. Beginning in the early 1970s, Mischel and his colleagues studied the ability of children to delay gratification. In what was later dubbed the marshmallow test, Mischel offered children as young as four years old the choice between a favorite treat, such as a cookie, and a less preferred one. But there was a hitch. The experimenter would leave the room, usually for fifteen minutes, and the child would have to wait for him to return to get the preferred treat. Or the child could ring a bell, and the experimenter would return immediately, but with the less preferred option.

Years later, when the kids were adolescents, Mischel found that the four-year-olds who could delay gratification were described by their parents as academically more focused than the kids who hadn't waited. The delayers also had fewer problems with frustration and were better able to resist temptation. Perhaps it wasn't so surprising that impatient kids became impatient teenagers.

Further work showed that the ability to delay gratification was a result of several cognitive factors. An individual's ability to cognitively transform the emotionally charged response of immediate gratification into something more abstract is perhaps the most important one. Mischel taught some of the children that, by thinking about a picture of their favored treat, they were better able to wait. In contrast, putting the real thing in the room with them made it impossible to wait.

Almost forty years later, the Cornell psychologist B. J. Casey performed the first brain imaging studies of Mischel's subjects.

Participants did an emotional version of the Go-NoGo task. While in the scanner, they were shown pictures of faces. One gender was deemed the "Go" stimulus, requiring a button-press, while the other gender was the "NoGo" stimulus, to which no button-press was required. In some conditions, the faces had a neutral expression, while other times the faces expressed happy or sad emotions. Amazingly, the participants who, forty years earlier, had been unable to delay gratification on Mischel's test had a higher false alarm rate to the emotional faces than the participants who had been able to delay gratification.

When Casey examined the patterns of brain activation, she found that a small region of the prefrontal cortex called the inferior frontal gyrus (IFG) was active during successful NoGo trials. Moreover, the participants who, as kids, had a better ability to delay gratification had higher activity in the IFG on the Go-NoGo task than their less patient peers. From these results, Casey traced a lifelong ability to delay gratification to the responsiveness of the IFG.

Casey's imaging results paralleled what we had found in the dogs. The brain region we had found in dogs that was associated with correct NoGo trials looked to be analogous to the region Casey had found in humans. To test the link, we needed a canine version of the marshmallow test.

We couldn't tell the dogs to wait, like Mischel had told the kids, because there was no simple way to teach the dogs that not waiting meant they would get a less desirable treat. As long as they got treats, it wasn't even clear that any of the dogs cared what the treat was. We would need a simplified marshmallow test.

Peter suggested something along the lines of putting a treat in front of the dog and making him wait to get it. We had all seen videos of dogs balancing treats on their noses until the owner gave the command to eat them. But none of the dogs in the project knew that trick. We would have to teach them.

To make it easier, we had each owner place her dog in the down position. Then we put a treat in a shallow cup, placed six feet in front of the dog. Peter had rigged a pulley system to the cup. If the dog broke from the down position, Peter simply hoisted the cup out of reach. The dog only got the treat when the owner commanded the dog to break position.

Our plan was to see how long each dog would stay in position before she broke on her own. The time a dog stayed would be an indication of her self-control. But we had troubles right from the start.

Kady never moved. She just put her head on her paws and looked at Patricia, waiting for further commands. After five minutes, Kady had fallen asleep. Only after Patricia coaxed her did Kady go after the treat.

A boisterous Portuguese water dog, appropriately named Tug, did the same thing as Kady, but for different reasons. Tug had been in the project for two years and had been one of the more difficult dogs to train to hold still in the MRI. He was only two years old when he started, so his youth was probably part of the reason, but he was also a high-energy dog. It was only because his owner, Jessa Fagan, had trained him well that he learned to do the MRI experiment. Jessa had no problem commanding Tug into the down position for the marshmallow test, but because of his high energy, I expected Tug would go for the treat immediately. And he wanted to. With great intensity, Tug stared at the cup on the floor. He glanced at Jessa, and then he started barking. But he never broke position.

Kady and Tug pointed to a larger problem. With the doggie-marshmallow test, we couldn't differentiate between self-control and training. It wouldn't be fair to penalize a dog who didn't know what to do. And even though Kady and Tug both stayed in position, their behaviors suggested very different reasons. Tug wanted to go for the treat but stayed in position, while expressing his frustration by barking. Kady either didn't care about the treat or was too inhibited to initiate any action without Patricia's command. Neither explanation required self-control.

Halfway through testing, we decided to scrub the marshmallow test. We would need something that tapped into self-control that wasn't so dependent on a dog's prior training.

Once again, we took a cue from the human development literature. This time, we went back further than Mischel, all the way to the grandfather of developmental psychology: Jean Piaget. Piaget is best known for his comprehensive theory of cognitive development in children, in which he formulated the idea that cognitive skills appeared in a specific sequence. In the sensorimotor stage, which begins at birth and continues until a child has the rudiments of language (at around two years old), infants learn about the nature of their environment by interacting with it. Around his or her first birthday, an infant learns that objects don't go away when they are out of view. Understanding this "object permanence" is a critical milestone. After this step is achieved, the game of peek-a-boo ceases to entertain, because the child is no longer surprised by the reappearance of his or her mother.

Peek-a-boo isn't a very precise test, so Piaget invented something a bit less animated. In the A-not-B test, two boxes are placed in front of the baby. The experimenter places a toy under box A, and then, after a brief delay, reveals the toy, much to the delight of the infant. The activity continues in this fashion for several trials, during which infants would often begin reaching toward the box to get the toy. Next, the experimenter makes a crucial switch and hides the toy under box B. Infants younger than about ten months continue to reach toward the A location, despite clearly watching the tester place the toy under the other box. By twelve months, almost all kids track to the correct location.

Failure on the A-not-B test indicates a disconnect between an infant's sensory world—seeing where the toy is placed—and his motor system—reaching toward the wrong location. To an observer, it is almost as if the child can't stop going habitually to A. This repeated error is called perseveration, and it has been linked to the immaturity of the frontal lobes. Comparative studies provide evidence of this phenomenon in other animals. Rhesus monkeys easily perform the task for hidden food, but if they have damage to the prefrontal cortex, they fail just like a nine-month-old infant.

The beauty of the A-not-B test is its simplicity. An experimenter doesn't need to teach an infant what to do and, by substituting food for toys, a wide range of animals can participate. In a landmark 2014 study, a consortium of animal researchers set out to test the performance of thirty-six species on the A-not-B test, including monkeys, apes, lemurs, birds, elephants, rodents, and dogs. On average, 89 percent of the dogs tracked to the B location and selected it on the first switch trial. This level of performance was comparable to that of the great apes (chimpanzee: 87 percent; bonobo: 100 percent; gorilla: 100 percent), but better than that of coyotes (29 percent) and many monkeys (capuchin:

86 percent; long-tailed macaque: 67 percent; squirrel monkey: 16 percent). Most of the birds performed poorly, except, perhaps, for some of the pigeons (55 percent). Surprisingly, none of the elephants could do the A-not-B task correctly.

When the researchers looked for patterns in the data, they found that the strongest predictor of performance on the A-not-B test was the size of the animal's brain. Elephants notwithstanding, bigger brains generally did better. Now, there was a bit of a confound here, as the bigger animals had bigger brains. When the researchers accounted for body size, the relationship between brain size and performance persisted, but not quite as strongly. Environmental variables, such as the percentage of fruit in the diet, or the size of a species' social group, were not correlated with A-not-B performance once brain size was taken into account.

Brain size alone could not explain all the differences. Relatively speaking, dogs do not have big brains, and yet they did as well as the primates on the A-not-B task. Coyotes did not do as well as dogs, even though dogs and coyotes are closely related and their brains are pretty much identical. And what about the fact that not all the members of a single species were able to do the task? Perhaps, as in Mischel's marshmallow test, some individuals were just better at it than others. And if that was true, then there might be a relationship between frontal lobe activity on a Go-NoGo task and performance on the A-not-B test, which meant we might have found our next experiment.

Months after we had completed scanning the dogs on the Go-NoGo test, we set up a version of the A-not-B test. Using baby gates, we cordoned off a section of a room, creating a chute about six feet wide and ten feet long, and we attached three buckets to the far end of the chute. The middle bucket would serve as a distractor and make it easier for us to tell if a dog was guessing. The left and right buckets would be the "A" and "B" locations. To make sure the dogs couldn't solve the task simply by scenting, we taped a treat behind each bucket. Pearl, the little golden retriever, was up first. With Pearl watching in great anticipation, one of my daughters, Helen, placed a treat in one of the buckets. Then, to avoid giving visual cues, she turned her back.

To show her that it was okay to eat it, Mark led Pearl to the bucket with the treat. Wagging all the way, she promptly lapped it up.

Satisfied that Pearl understood the game, we repeated the trials two more times, but now, Mark released Pearl from the end of the chute and let her go on her own to bucket A. On the fourth trial, Helen at first placed the treat in bucket A, and then moved it to bucket B. From behind, I could see Pearl's head tracking Helen's movements. Pearl knew where the treat was.

Mark released her.

Pearl trotted to the middle bucket. Seeing nothing inside, she then went back to bucket A, where she was similarly disappointed.

I looked at Peter and shrugged. By the classic definition, Pearl had just failed the A-not-B test. It was a soft failure, because she almost made it to bucket B by going to the middle, but a failure nonetheless.

Anticipating such mistakes, we had decided that the dogs would get multiple chances to make the switch. Their performance would be scored based on the number of attempts. Pearl was no dummy. On the second try, she went straight to the B location. The dogs exhibited a striking range of ability on the

A-not-B test. Zen and Big Jack tracked to the B location on the first switch trial, while Kady and Eddie did the worst, requiring eleven attempts before making the switch.

Dogs with higher false alarm rates on the Go-NoGo test tended to take more trials to switch on the A-not-B test. Because the tests both measured aspects of self-control, but in different ways, the concordance between the results showed that we were picking up a general trait of each dog.

The results of the experiment could be summed up by a simple chain: low frontal activity ⟶ high false alarm rate on Go-NoGo ⟶ more perseveration on A-not-B. The relationships were correlational, so we couldn't directly infer causality, but the data about frontal lobe function in humans and other animals made a strong case for the pivotal role of the frontal lobe in dogs, too. Echoing the Mischel/Casey experiments, dogs with a more active frontal cortex did better on cognitive tasks requiring self-control than dogs with a less active frontal cortex. So the answer to the question of what it's like for a dog to exert self-control is within our own brains. Whether it's pushing away a second helping of dessert, or resisting an impulse purchase, or kicking an addiction, everyone knows what it feels like to exert self-control. The brain data suggested that a dog's experience was very much the same.

The dog results were important for two reasons.

On a practical level, dog owners resort to a dizzying array of tactics to keep their pets from swiping food off of counters, digging up flower beds, and chewing up all manner of clothing. Dogs will be dogs, and these are things dogs like to do. And yet, dogs are not all equal. I have no problem letting Callie have free run of the house while I am at work, but there is no way I could ever let her off-leash in an unfenced area. No amount of self-control could quell her terrier-drive to chase small animals—or, more dangerously, cars. Her privileges don't depend on her being a dog, but on her individual nature. The other dogs in the house get different privileges based on their abilities, too.

Second, the link between frontal lobe activity and a dog's behavior parallel a similar link that has been observed in humans. Analogous regions in dog and human brains appear to serve analogous functions. This is important, because analogous structure-function relationships provide a pathway for answering the question of what it's like to be a dog, or any other animal. I suspected that when analogous brain structures were active in an animal, they were having analogous subjective experiences to us. The fact that we found individual differences in dogs, just like there are in humans, made this even more likely. After all, the field of human neuroscience was shifting in the direction of personalized medicine, where understanding individual differences in biology is becoming paramount.

The Dog Project had moved beyond generalities of canine brain function. We began to focus on the subtle differences between individual dogs and what those differences said about their individual experiences. But the same approach to individuality within a species could also be applied between species. In going beyond the dog, one moves from descriptions of what it's like to be Kady or Libby to descriptions of what it's like to be a dog or some other four-legged carnivore, or a two-legged one.

Resolving the tension between individuality and species requires a deeper examination of the relationship between structure and function in the brain. What brain structures do species have in common with each other, and how variable are they within a single species? Before tackling these questions, we must first grapple with an even more basic one: *What, exactly, is a brain for?*

Chapter 3

Why a Brain?

A brain is an expensive piece of machinery to maintain. In humans, the brain is a mere 2 percent of the body's weight, but it commands 20 percent of the blood flowing from the heart and uses 20 percent of the oxygen we breathe. It is so sensitive that it cannot withstand even the slightest interruption in blood flow. Quick drops in blood pressure can result in immediate unconsciousness, and in a brain that goes without blood and oxygen for more than about five minutes, irreversible damage occurs. Ten minutes without oxygen, and you are dead.

Imagine driving a car that was this temperamental. Who would want a machine that, even with the slightest lapse in maintenance, would cease to run, never to be fixed? For such a fickle piece of biology, the advantages of having a brain must far outweigh its costs. So what, exactly, is a brain for?

From a naïve Darwinian perspective, a brain allows animals to survive and reproduce, but that does not explain why some animals have bigger brains than others, or why humans have such large frontal lobes. These differences in brain structure underlie the differences in function between species. The challenge is to decipher this structure-function relationship. This is a daunting task considering that the human brain contains at least 80 billion neurons, although the exact number is hard to know. Judging by its relative size, it is estimated that a dog brain probably contains at least 5 billion.

And somewhere, in that tangle of neurons, was the key to understanding the mental experiences of other animals. Some researchers and philosophers had said it was impossible to understand an animal's mental life. I thought otherwise. The Dog Project pointed to similarities of both structure and function between the brains of dogs and humans. But functional MRI wasn't the whole story. While fMRI was great at revealing changes related to neuronal activity, its resolution was limited. A lot was happening below the surface of those functional images. To get to the physical basis for the experience of being a dog, or any other animal, including a human, we would need to go deeper into the architecture of the brain.

Our understanding of the brain has changed dramatically in the past one hundred years. Although I took issue with the assertion that we could never know what it was like to be a dog, I also had to acknowledge that, viewed through the history of neuroscience, the prospects may have looked pretty dismal back in the 1970s. Much has changed. Not only has there been an explosion in neuroscience information, but theories of brain function have evolved in parallel with technological revolutions.

Technology dictates what can be measured, but technology also serves as a metaphor for interpreting biological systems. It has always been that way. As far as brain function, three themes stand out: electrical switches as a metaphor for stimulus-response relationships; early computers as a metaphor for symbolic manipulation; and the integrated circuit as a metaphor for neural nets. Looking at the origin of these themes helps situate current theories of brain function and leads to a few general principles concerning what brains do. From there, one may consider what a dog brain does and how that is different from what a human brain does.

Neuroscience began in earnest in the early twentieth century. The electrical revolution was in full swing, too. Thomas Edison had patented his lightbulb design in 1879, and by 1900, Guglielmo Marconi was testing his radio. Even before Marconi, Alexander Popov had invented a lightning detector based on radio waves. St. Petersburg, Russia, where Popov was based, was a hotbed of intellectual activity. It was also the home of Ivan Pavlov, a pioneer in the study of reflexive behavior. Although we don't know if Pavlov knew Popov, the use of electrical devices figured prominently in Pavlov's work. Crucially, Pavlov demonstrated that reflexes weren't just innate. They could be learned through a process he called conditioning. The analogy to electrical circuits was unmistakable. Like an electrical switchboard, reflexes could be reconfigured.

Pavlov's discovery of conditioned responses dominated psychology for the next fifty years. In 1911, the psychologist Edward Thorndike published his law of effect, which stated that actions that preceded enjoyable responses were more likely to be repeated. From this simple observation, another psychologist, B. F. Skinner, would eventually build a theory of operant conditioning.

Although it was well known that the brain operated on some form of electrochemical basis, the line of thought from

Pavlov to Thorndike to Skinner was fundamentally mechanistic. The brain was treated as a black box: unobservable and irrelevant to the study of behavior. The first step past the Skinnerian black box came in the 1950s, when scientists began looking to the brain, particularly how it stored information, for new clues. This view of the brain was more sophisticated than the stimulus-response relationships of the previous era.

In large part, the newfound interest in the brain was driven by the invention of computers, which led psychologists to look at brains as biological computing devices. This new "cognitive psychology" emphasized the representation of knowledge in the brain and how this information was manipulated. And yet, with a heavy emphasis on the software that constituted the mind, the brain once again took a backseat as just some biological hardware. Many researchers fantasized about the day we could do away with the brain entirely and just download the software into a computer. The possibility of major breakthroughs in our understanding of the brain as well as artificial intelligence kept the cognitivist movement alive for the next thirty years.

By the mid-1970s, though, a growing contingent of academics began to realize that brains did not store things in the same way computers did. Brains didn't have separate memory banks and central processing units (CPUs) like computers. Without a basic understanding of how knowledge accumulated in the brain, researchers were at an impasse in separating the software (the mind) from the hardware (the brain). Thomas Nagel's essay merely fanned the flames of this discontent. He said, in effect, that deconstructing the brain would never explain the mind. Although Nagel was reacting to the apparent futility of reductionist approaches to the mind, his essay cleaved academics into believers and nonbelievers of the utility of neuroscience, a split that persists to the present day.

In defense of the biological approach to the mind, a new breed of scientists emerged who flipped the problem on its head. Instead of looking for computer-like functions in the brain, they took inspiration from neuroscience and began designing computer algorithms that mimicked how the brain worked. The first thing they noticed was the high degree of parallelism in the brain, with billions of neurons, all working at the same time. Such massively parallel computing is very different from a computer, in which the CPU executes instructions sequentially. The new "connectionists" showed that simple networks of neuron-like automatons could perform surprisingly complex tasks. Moreover, these neural nets could learn on their own without the godlike hand of a programmer.

These early neural-net models were powerful in their ability to perform humanlike tasks, such as recognizing handwritten letters or beating humans at backgammon. The growth in neural nets paralleled the technological explosion in integrated circuits, and it wasn't long before people started making neural-net chips. Now, with the vast availability of information and limitless computing power, neural nets have been joined with artificial intelligence (AI) algorithms, leading to a new hybrid called deep learning. But neural nets are still just input-output devices. They model some type of input from the environment, transform it, and produce an output.

The input-output analogy of brain function seems to make sense. We constantly take in information, think about it, and sometimes act upon it. But the analogy is fundamentally backward. Brains did not arise to process information. They evolved to control the movements of creatures. Indeed, all animals with muscles have nervous systems, and all animals with nervous systems have muscles. The one-to-one relationship between nervous systems and muscles leads to one inescapable conclusion and the first principle of brains:

Animals have brains because they need to do things.

What an animal needs to do, though, depends both on its physical form and on the environment in which it is situated. Although brains do process information, information processing is necessary only to the extent that it facilitates action. Moreover, animals have control over the information they process, which is called active perception. Buried in this tight relationship between brain and body is the animal's mind.

To see how this came about, we must look to the origin of animals and their nervous systems.

Although the earliest life-forms appeared 4 billion years ago, not long after the planet formed, it took more than 3 billion years for the first animals to show up. Prior to this, there probably wasn't enough oxygen in the atmosphere to support complex life-forms. But about 600 million years ago (mya), enough oxygen had accumulated and life exploded (in what was called the Cambrian explosion). Multicellular organisms rapidly increased in complexity, and the first creatures that we would recognize as animals appeared. They looked very much like modern jellyfish.

Jellyfish have nervous systems, and they have a ring of muscles for propulsion. But a jellyfish has no brain. Jellyfish and their relatives, which belong to the phylum Cnidaria, have nerve nets. The difference between a brain and a nerve net is in the degree of centralization. Nerve nets are rather simple nervous systems and are characteristic of animals that don't have complex movements.

Even though jellyfish don't have a centralized command center, their nerve nets are capable of a surprising range of behaviors. Predatory jellyfish are able to sense nearby prey when the prey comes in contact with their trailing tendrils, called nematocysts. Contact initiates a reflex through the nerve net, resulting in the firing of the nematocyst and harpooning of the prey. There is no consciousness involved in the act. Indeed, there is no "it" to be aware of its actions, because there is no centralized system to keep track of what everything is doing. Jellyfish are the zombies of the ocean.

A rather obvious feature of jellyfish is that they are radially symmetric. They are living tubes. Because of this geometric shape, their nerve nets are also distributed in a tubular arrangement.

In the evolution of the nervous system, the next big change came when animals broke from radial symmetry to left-right symmetry. Imagine taking a tube and flattening it lengthwise. Now it is no longer radially symmetric. Instead, it is symmetric only across the long axis. In essence, we have created left and right, and front and back. Taxonomically speaking, these were the first *bilaterians*: that is, they had bilateral symmetry. In their simplest form, they looked like flatworms.

As the bilaterian bodies flattened out, so did their nerve nets. What was once radially symmetric became compressed into two cords running down the left and right sides, connected in between by a sparse transverse net. As in jellyfish, one end of the tube was not the same as the other. These new creatures had an identifiable head and tail. The bilaterian nervous system shows what happens when neurons are compressed into closer proximity. They have the opportunity to form more connections with each other. And more connections mean more complex calculations. In a bilaterian nervous system, one of the most important tasks is to coordinate the actions of the left and right sides. You couldn't very well have the left and right sides moving in opposition. So the first bilaterians had the first primitive central controllers.

Even at such an early stage of animal evolution, you can see how intertwined nervous systems are with movement. Just wriggling in one direction requires a high degree of coordination. Indeed, coordination is so important that large chunks of the human spinal cord and brainstem are still devoted to it—a reminder of our inner worms.

We don't need to trace the entire course of evolution from worms to present-day animals to understand a few key principles of brain function. First, although evolution doesn't proceed in a linear fashion, left-right coordination is so important to all animals that once evolution "solved" it, it seems that every animal subsequently inherited these solutions. Second, once the nervous system gained control of the body, the next crucial function of a nervous system was deciding what to do. This was decision-making, and for this, animals needed brains. Indeed, this returns us to the first principle: *Animals have brains because they need to do things.*

Evolution is the single guiding principle in all of biology, but it is not easy to appreciate how evolution resulted in the modern brain. In the strictest sense, as Charles Darwin realized, evolution acted on the whole organism. An animal must survive long enough to reproduce, and then it must reproduce. Individual body parts, like the brain, evolve only because changes arise that give an animal either a survival advantage or a reproductive advantage. So even though we can talk about the evolution of the brain, we must consider how changes in brain structures make an animal better adapted to its environment, and not simply how the changes made the brain more like ours. These considerations lead to the next principle of brain function:

Animals have brains to tailor their actions to their environment.

In other words, animals do not exist in isolation. They are embedded in the world around them, and part of the function of a brain is to link the external world to the animal's decision-making systems, and ultimately, its body.

B eyond worms, animals really started to get interesting when they developed a backbone. The first vertebrates appeared 500 million years ago. Initially, they didn't look too different from worms, except that they were quite a bit bigger, and because of their size, they needed something more substantial than soft tissue for structural support. Enter the notochord—a tough thickening that ran the length of these animals. As they got bigger, the control and coordination problems necessitated more complex nervous systems, which led to further consolidation and centralization.

Our best living models for these creatures are the lampreys and hagfishes, which are ocean bottom-feeders with no jaws. Depending on your point of view of such things, hagfish and lampreys are either the most fascinating or scary-looking creatures in the ocean. Hagfish are fairly innocuous and spend most of their lives wriggling through muck on the ocean floor. When threatened, they secrete a thick slime. Lampreys, though, look like something out of a sci-fi movie. (The sandworms in *Star Wars* and *Dune* took their inspiration from lampreys.) Lampreys attach themselves to larger fish with their large mouth suckers. Behind the suckers, rings of teeth gnaw away at the flesh of the unfortunate host. Because fish had not yet appeared when the first lampreys swam the oceans, the teeth must have been a later adaptation.

The early jawless vertebrates had what could be considered the first recognizable brain. A swelling at the head-end of their spinal cords contained all the basic parts found in every vertebrate brain: an olfactory bulb, a primitive cortex for decision-making, a region for sensory processing, and an area for coordination and regulation of life-maintaining functions.

These first vertebrates had the bodies and the brains to control their behavior to an extent that invertebrates couldn't. But with all these different animals swimming the oceans, the competition had heated up too. Reflexes wouldn't cut it. To survive in these ancient oceans, an animal had to make better choices than its competitors. Animals that could adapt their behaviors had a distinct advantage over those that had only hardwired motor programs. The need for flexible behavior leads to the third principle of brains:

Animals have brains so they can learn.

In actuality, even animals with simple nervous systems can learn, but there are different degrees of learning. A stimulus-response association is a form of learning, and you don't need much more than a few neurons for that. But the type of learning that full-fledged brains accomplish is much deeper. A sentient creature can and must learn that the environment contains both good and bad outcomes—mostly the latter. Survival and reproduction depend largely on making a long series of good choices while avoiding the bad ones that can kill you. There are no second chances. So how does an animal learn from these experiences without actually getting killed?

The answer is that brains have evolved to do more than take in information and act on it. A sophisticated brain constantly runs simulations of possible actions and outcomes, like you do when you are playing chess. As the behavioral repertoire of vertebrates increased, the complexity of their brains had to increase accordingly. Some of this was to keep pace with the complexity of their actions, but mostly the increase in brain size reflected the growing need to outwit competitors in the Darwinian game of life. And while learning is backward-looking, the need to look forward and anticipate different possibilities leads to the fourth brain principle:

Brains simulate possible actions and future outcomes so as to make the best possible decision for the situation at hand.

After another 80 million years of evolution, fish appeared. These were the cartilaginous fishes, which included sharks and rays. And in the Darwinian push for survival of the fittest, fish became bigger and stronger, necessitating stronger skeletons. By 400 mya, the oceans were filled with a wide range of both cartilaginous fish and their bony cousins. The evolution of bone was a key development that allowed the appearance of more complex body shapes. Fins appeared in more locations, increasing both speed and maneuverability. Finally, about 390 mya, in perhaps the most important innovation in vertebrate evolution, some of the fins became sturdy enough to support weight on land. Enter the tetrapods.

The earliest tetrapods, which looked like salamanders, lived mostly in water, occasionally venturing onto shore. No doubt they found a landscape rich with plant life. With all this food to themselves, it is easy to see how these early amphibians gained a huge survival advantage over their water-bound peers.

The amphibians laid their eggs in water because, lacking a protective covering, the eggs could not survive on land. As a result, evolution again found a niche—an advantage for tetrapods that produced eggs able to endure on land, out of reach of the ocean-dwellers that liked to dine on them. About 320 mya, eggs became tough enough to survive solely on land, and the animals that produced them, called sauropsids, evolved into the reptiles and birds. In time, the sauropsids ruled the Earth. They were a varied bunch with a high degree of species diversity. The party ended rather suddenly 250 mya in a mass extinction (the Permian-Triassic extinction). The cause is unknown, and theories range from massive meteor showers to volcanoes to a runaway greenhouse effect. It took almost 10 million years for life to recover.

The surviving sauropsids eventually became the crocodiles and dinosaurs. In fact, the dinosaurs got a boost 200 mya, during another mass extinction that wiped out much of their competition (the Triassic-Jurassic extinction). What was left of the non-dinosaurs on land evolved to become smaller and cleverer to outwit the now massive dinosaurs. These became the predecessors to the mammals.

The earliest ancestors of mammals, the cynodonts, still laid eggs like reptiles and birds. The cynodonts had been around since the Permian-Triassic extinction and looked like a cross between a rat and a lizard. Their legs were more vertical than the reptiles', which increased the cynodonts' mobility. The cynodonts may have even had some basic thermoregulatory mechanisms. If so, they were the first warm-blooded animals. The larger cynodonts had enough body mass to retain heat, but the smaller ones may have had fur to keep warm in cold temperatures. In time, a new reproductive strategy emerged that allowed their eggs to incubate internally, which protected them from being eaten before they were born. This new branch of mammals, called the therians, gave birth to live young. They are called the "crown mammals" because all current living mammals can be traced back to them.

The dinosaurs might very well still be here were it not for the impact of a large asteroid about 66 mya, which resulted in the extinction of all the dinosaurs except birds. The Cretaceous-Tertiary event was the fifth, and most recent, mass extinction, and even though all species were hit hard, the mammals gained the upper hand during the recovery period. With the loss of the dinosaurs, the mammals rapidly diversified to fill ecological niches left behind.

And the mammals got bigger, which brings us back to brains.

B ig bodies have big brains. This may seem obvious, but the origins and implications of this fact have been points of intense debate for over one hundred years. To understand the animal mind from a biological perspective, we must first account for the differences in brain size and what, if anything, all that extra neural tissue does for big animals.

In 1973, Harry Jerison, a psychologist, proposed a simple rule to explain the difference in brain size across species. He wrote, "The amount of neural tissue controlling a particular function is appropriate to the amount of information processing involved in performing the function." He called it the *principle of proper mass*. Because biological systems required constant sources of energy, Jerison reasoned, a chunk of brain would evolve to be only as big as necessary to accomplish its purpose. Any more would be wasteful. Flipping the logic around, the size of a brain structure would tell us something about the amount of work it did relative to other parts of the brain.

By Jerison's logic, larger brains must therefore process more information than smaller brains. Why should that be?

The first clue comes from both the geometry of animal bodies and the mathematical relationship between brain weight and body weight. Small birds have a brain-to-body ratio of 1:10, dogs and cats are about 1:100, elephants about 1:500, and blue whales have a brain-to-body ratio of approximately 1:14,000. So even though big animals have big brains, the increase in brain size becomes smaller as the animal gets bigger. The weight of the brain turns out to be roughly proportional to the weight of the body, raised to the exponent 2/3.

The exponent 2/3 is important because basic geometry dictates that the surface area of a body is proportional to its volume raised to the 2/3 power. The mathematical relationship didn't arise because bigger animals had more muscles and therefore more to control. Insects, for example, have about the same number of muscles as we do. Instead, as Jerison astutely realized, more surface area meant more sensory inputs coming from the skin, which had to be processed in the brain.

Scientists love mathematical rules that explain biological phenomena, but such rules are most useful as guiding principles and not as inviolate laws. For all such rules, there are always exceptions, with humans being a notable one. Our brains are much bigger than would be predicted by the surface-area rule. Moreover, the surface-area explanation fails to take account of other sensory systems, notably vision, the importance of which varies tremendously across species.

A new measure, the encephalization quotient (EQ), emerged as a way to measure brain size after accounting for the effect of body size. So even though elephants have huge brains, the EQ can tell us whether it is huge for its body size. Jerison defined the average EQ for mammals to be 1. If an animal species had an EQ greater than 1, it had a big brain for its body size, while an EQ less than 1 meant comparatively little brain power. Cats were average, with an EQ of 1, while dogs fared slightly better at 1.2. Monkeys, chimpanzees, and elephants had EQs around 2, while the bottlenose dolphin clocked in around 4. Humans sat atop the brain heap with an EQ of 7.

It is tempting to conclude that a higher EQ means a smarter animal, but only in the broadest sense does that seem to be true. If the ability to form abstract concepts and have a spoken language is equated with intelligence, then of course humans come out on top, and it is all too easy to conclude that our bigger-than-average brains are responsible. But even within humans, the EQ-intelligence relationship breaks down. Take two people with the same size brain. If one person weighed 150 pounds, and the other weighed 250 pounds, the first would have an EQ of 7 while the heavier one would have an EQ of 5. There is no evidence that losing weight makes a person smarter.

In recent years, EQ has come under fire because it assumes that brain matter is the same for all animals, which may not be true. Since 2006, Suzana Herculano-Houzel, a Brazilian neuroscientist, has been working on how to measure the number of neurons in a brain. Until her work, there hadn't been any reliable way to do this. It had all been based on randomly sampled cores from different regions that were extrapolated to the whole brain. Herculano-Houzel figured out how to dissolve whole brains into a soup in which she could separate the neurons from everything else. When she counted the number of neurons in the brains of different animals, she found that humans aren't that special. Although humans do have a large number of neurons, about 86 billion, the number is appropriate for our body size when compared to other primates. The big difference is between primates and all the other mammals. Primate neurons appear to be smaller, and because of their smaller size, they can be packed more densely into a brain of a given volume. Herculano-Houzel has argued that it is the number of neurons, especially in the cerebral cortex, and not EQ, that is the critical determinant of intelligence.

Whether it is brain volume or neuron number, these are still overall measures of brains, which may be about as informative as a person's height or weight. To understand an animal's subjective experience, we must go deeper into the organization of the brain.

Because overall brain size doesn't explain much, we must look to other variables, and the next variable to consider is the size of specific parts of the brain. Individual brain regions should still follow the principle of proper mass, with larger regions performing more information processing, which should tell us something about an animal's internal experiences. And because neurons are expensive to run, the size of a brain region should say something about the importance of its function to the animal.

Before delving into individual brain regions, though, we need to clarify an ambiguity about size. There are three ways to measure size. The first, and most straightforward, is the absolute size of a region, which is given by its volume. The second is the proportional size of a brain region. Proportional size is just the ratio of the volume of a brain region to the overall brain volume. Proportional size is quite interesting because each division of the brain has its own relationship to overall brain size. For example, as brains got bigger, the cortex occupied a greater and greater proportion. Other regions, like the cerebellum and the brainstem, got bigger too, but not quite as fast as the cortex did. These proportional sizes of cortex, cerebellum, and brainstem are surprisingly consistent across species, especially mammals. One theory suggested that the major divisions of the brain evolved in a coordinated fashion. This makes sense. Because a brain is so interconnected, anything that affects one region would affect others.

But if brain regions developed in concert, then how could evolution act to augment or diminish specific functions? Such differentiation is at the heart of what makes the brain of a cat different from the brain of a dog, or that of a human different from that of a chimp. This paradox leads to the third way of measuring the size of a brain region.

Whereas proportional size is a ratio of a region to total brain volume, relative size is the ratio of regions to each other. Dogs have large olfactory bulbs, but are they large for their brains? In dogs, the olfactory bulb occupies about 0.3 percent of the total brain volume. If we include the surrounding neural tissue (the olfactory tract and stria), the proportion rises to 2 percent of the brain. In humans, these ratios are 0.01 percent and 0.03 percent, respectively, but the proportional volumes could be small just because other parts of the cortex got much bigger. We really need to know the volume of the olfactory system relative to some other sensory system, like vision. Only then can we compare the dog's sense of smell to the human sense of smell. If evolution acted on the relative sizes of regions, then perhaps brains evolved as a mosaic of individual pieces, each under its own evolutionary pressures.

Another example of mosaic evolution can be found in the relative importance of auditory and visual information. In the auditory domain, sound, in the form of pressure waves, enters the ear, where the waves are transformed into vibrations of the small bones in the middle ear. Hairs sticking out of specialized neurons in the inner ear transform the vibrations into electrical impulses that travel up the auditory nerve to the brainstem. As the auditory signals make their way toward the brain, they pass through a series of structures, the most prominent of which is called the inferior colliculus. The left and right colliculi form a pair of bulges on the back of the brainstem. As you might expect, if there is an inferior colliculus, there is a superior colliculus, which sits atop the lower one and is the analogous structure for receiving visual information. Anatomists had long recognized that the relative sizes of the inferior and superior colliculi correlated with the relative importance of auditory and visual information for an animal. Bats and dolphins, which use echolocation, have larger inferior colliculi than superior ones, while animals that rely more on vision, including many primates, have larger superior colliculi.

One of the most compelling examples of the relationship between relative size and function is found in bird brains. The hippocampus is a structure that is nestled in the crook between the cortex and the brainstem. In mammals, it forms an upsweeping curve along the inside of the temporal lobe. In birds, the hippocampus lies along the top of their brains. In a classic study, John Krebs, a zoologist at the University of Oxford, measured the relative size of the hippocampus in birds that store their food, such as crows, and those that do not, such as finches. After controlling for both the body size and overall brain size, Krebs found that the species that store their food have larger hippocampi than those that do not.

The hippocampus has long been known to be the critical structure for memory formation. In the 1950s, Henry Molaison, an epilepsy patient, had both his hippocampi scooped out in an effort to cure the disorder. In that sense, the operation was a success. In another sense, H.M., as he was known until his death, became the most famous neurologic patient in history, because he could no longer form new memories. In animals, the hippocampus is particularly important for spatial memory—remembering where things are. So it makes perfect sense that birds that store food for later consumption need to have brains that devote more territory to spatial memory.

The aforementioned examples are the ones most frequently trotted out to support theories of mosaic brain evolution. They are powerful examples and illustrate foundational relationships between the relative sizes of brain regions and their functions. But most other cases provide considerably weaker evidence. Variations in the size of other brain regions tend to follow the general relationship to overall brain size. As animals got bigger, so did their brains, along with all their constituent parts.

But, as in life, size isn't everything. It's who you're connected to.

Neurons get the lion's share of attention, but the gray matter of the cortex, which contains the cell bodies of neurons, is a scant 3 mm thick. Most of the brain is made up of other stuff. Glial cells, which support neurons structurally and metabolically, make up a lot of this stuff. Cerebral spinal fluid, or CSF, provides a cushion by keeping the brain floating in water. And then there is the white matter, which occupies large chunks of the brain. The whiteness comes from a waxy substance called myelin that acts as an electrochemical insulator around the axons of the neurons. The actual synapses, where neurons transmit information to each other, are typically located in the gray matter, but the white matter allows neurons to communicate with each other over long distances. This can be from one part of the brain to another, or between the brain and the spinal cord. Indeed, the axons within the white matter of the spinal cord can be up to a meter long.

Jerison studied gray matter because that was where the neurons were. Until recently, few neuroscientists cared about the white matter, and it was a surprise when Kechen Zhang and Terrence Sejnowski, theoretical neuroscientists at the Salk Institute, discovered a powerful relationship between white and gray matter. Zhang and Sejnowski measured the gray and white matter volume in the brains of fifty-nine different mammals. The animals ranged in size from small pygmy shrews to elephants and pilot whales. When they plotted the gray and white matter volumes on a logarithmic scale, the values lay on a straight line. On this type of graph, the slope gives the scaling exponent. They found that white matter volume was equal to the gray matter volume raised to the power 1.23.

This power is interesting for two reasons.

First, the exponent is greater than 1, which means that white matter volume increases faster than gray matter volume. As brains get bigger, white matter takes up more space. This makes sense because the more neurons a brain has, the more connections they have to make with each other. The surface of the cortex is, for the most part, a sheet of neurons, so as the brain gets bigger, the sheet has more surface area. If all the neurons were connected to each other, then we would expect the number of connections to increase as the square of the number of neurons. But it doesn't.

Second, the exponent is less than 2, which means that as gray matter increases, white matter increases at a faster rate, but not a rate that allows neurons to stay fully connected with each other. Failure to keep apace with full connectivity means that bigger brains break into discrete chunks. In other words:

As brains get bigger, they become more modularized.

Zhang and Sejnowski realized that the relationship between gray and white matter volume could be explained by a simple principle in which the brain minimized long-distance connections. Fibers that traveled long distances would occupy a lot of space and introduce conduction delays. White matter tracts are like the highways that a fleet of delivery trucks uses—necessary, but expensive to maintain. Imagine you had to ship a bunch of goods across the country. You could send each package individually from a central location, or you could accumulate them in regional warehouses and send them locally. The long-distance costs are minimized by consolidating shipments to the regions.

In what might seem an otherwise obscure mathematical relationship lay a deep truth about brain organization. Prior to Zhang and Sejnowski, scientists had argued over why animals' brains looked different from each other. The principle of proper mass implied that the size of a brain region was linked to the amount work it did, either proportionally or relatively. Zhang and Sejnowski showed that there were costs associated with size. As parts got bigger, the amount of brain devoted to connecting the parts grew even faster. This led to a paradox. Evolution had pushed brains to become centralized controllers of the body, but as animals got bigger, their brains fractionated into more discrete pieces. Instead of a unitary thing, a modern large brain was a collection of semiautonomous modules.

The relationship between white matter volume and gray matter volume is foundational to understanding the brain, but the relationship is still about size. It does not explain why a dog's brain looks different from that of a rhesus monkey, though both weigh about 100 grams. To know how a dog's brain makes him a dog and not a monkey, we need to look deeper into the specific arrangement of the white and gray matter pieces. We need a detailed map of how things are connected to each other.

The challenge of figuring out how all the parts of the brain are linked together is like trying to figure out how a country's economy works from outer space. Imagine you were parked in orbit 250 miles above the Earth's surface, living on the International Space Station. How could you learn about the United States? You would probably start with the obvious landmarks: oceans, mountains, rivers, cities. That would tell you where the action was but not much else. If you had a keen eye for detail, you might notice the highways and how things moved between the hubs of activity. Eventually, you could piece together a theory of how the country works.

Neuroscience of the late twentieth century had focused on the landmarks, like the size of different brain regions, or what caused them to be active. But the twenty-first-century paradigm was all about mapping the highways, and the members of this generation of neuroscientists were the new cartographers of the brain. This was the study of *connectomics*.

More than just mapmaking, connectomics held the possibility of finally getting into the animal mind. All of those connections between neurons in different brain regions serve an important function. The connections coordinate activity, and it is only through this coordination that animals can be conscious of their surroundings and what they are doing. The pattern of connections provides the roadmap to the mind. Just like the United States and Canada have different roadmaps, so do dogs and monkeys. To know what it is like to be a dog, we need to look at the dog's roadmap.

Brain connectivity is so tightly linked to mental states that physicians have a name for disorders of connectivity: disconnection syndromes. When brain regions become disconnected from each other, they continue to function in isolation, and this leads to a variety of well-known neurologic conditions. The left and right hemispheres, for example, are pretty much complete brains in their own right. The split-brain experiments of the 1950s had shown that each hemisphere could process information and control the opposite side of the body in an appropriate manner. But without a corpus callosum to connect the hemispheres, the person could not explain why one side of the body was doing one thing and the other side another. The loss of connectivity diminished consciousness. While split-brain is a surgically induced condition, other problems occur as a result of strokes or traumatic injuries. A stroke to the set of fibers connecting the region responsible for processing speech to the region for producing speech results in a disconnection syndrome called conduction aphasia. Patients with this syndrome can speak fluently, but because the region that processes incoming language has been disconnected, they can't monitor what they are saying, and the result is a stream of gibberish.

Traumatic brain injury, such as what happens during the sudden deceleration of a car wreck, causes widespread damage to white matter throughout the brain. If the damage is severe enough, the cortex can become disconnected from the brain-stem. And because the brainstem contains clusters of cells responsible for regulating wakefulness, disconnection from these cells results in coma. With time, patients can recover from such injuries, but the healing process does not proceed at the same rate throughout the brain. Some pathways can become functionally connected again while others remain stunned. When this happens, even the slightest sensory stimulation can set off a flurry of cortical activity. The cortical activity, in turn, can manifest as wild behavior, screaming, or running around without purpose. The patient often shows little awareness of what he or she is doing. In the past, such behavior was treated with heavy sedation. Now, thanks to a better understanding of connectomics, physicians have learned to control sensory stimulation to patients emerging from comas, instead of using tranquilizers, and this approach results in faster recovery.

The treatment of brain injuries says a lot about what it's like to be an animal, because brain injuries reveal the link between coordinated electrical activity and consciousness. Indeed, consciousness is nothing but coordinated electrical activity. This also makes clear that consciousness is a continuum of awareness that fluctuates with physiologic state and environmental stimuli. Even in healthy human brains, consciousness can range from being minimally aware of one's surroundings, as during sleep, to the hyper-focused state necessary for goal-directed actions, such as performing brain surgery. The difference lies in the degree of coordination of electrical activity. When brain parts become disconnected from each other they cannot coordinate their activity, and the result is a disorder of consciousness.

Although consciousness fluctuates in all animals, the pattern of connections in an animal's brain provides the roadmap for the level of consciousness that is possible. Somewhere between the nerve net of the jellyfish and the human cerebral cortex is a connectivity map of sufficient complexity to give rise to the key domains of consciousness: perception, emotion, movement, memory, and communication. And beyond these lay domains of consciousness that transcend the individual, such as an awareness that other animals have internal mental lives too.

Perhaps more than anything else, higher levels of consciousness depend on memory. Only by keeping track of past events can an individual maintain a sense of self through time. Memory lets a person wake up each morning with the feeling that he or she is the same person as the night before. There is no single place in the brain for memories. They are distributed throughout the cortex, and it is only through the coordination of activity that memories can be recalled. Anything that interferes with this process, such as Alzheimer's disease, necessarily disrupts the sense of self. As the brain deteriorates, neurons and their connections with each other disappear, and with them, memories. Subjective experience literally collapses.

Although animals are not known to get Alzheimer's, there have been circumstances in which their memory systems have been destroyed by environmental toxins. In one case, humans have been afflicted by the same toxin, and the parallels between the human and animal syndromes are striking. Disorders of memory, like disorders of consciousness, say a lot about what it's like to be a person, or, in the case of an animal, being that animal.

Chapter 4

Seizing Sea Lions

In late 1987, as the holiday season was ramping up, Canadian health authorities found themselves in the midst of an epidemic like nothing before. Hundreds of people were flooding Montreal emergency rooms. All of them had nausea and vomiting. While that suggested an outbreak of some foodborne illness, many of the patients quickly became confused, which is not a typical symptom of gastroenteritis. The confusion was severe. Several patients developed seizures and descended into a coma. Four of them never woke up and were dead within a week. Those who finally did emerge from their comas were never the same again, having lost the ability to form new memories.

Between November 4 and December 5, 1987, 107 people had fallen ill. It didn't take long for health officials to figure it out. All of the patients had eaten mussels. And all of the offending mussels had come from a river estuary on Prince Edward Island, just north of Nova Scotia. To confirm that the mussels were, in fact, the cause of the illness, scientists injected some mice with extracts from the mussels. Within ten minutes, the mice began furiously scratching themselves. They became increasingly uncoordinated. All were dead within an hour.

None of the mussels contained environmental toxins. Instead, toxicological analysis showed high levels of domoic acid. Not much was known about domoic acid, but chemically it behaved much like the neurotransmitter glutamate. Glutamate is the most common neurotransmitter in the brain, and when released, it excites other neurons. Too much glutamate results in overstimulation and seizures. When excess glutamate pours out of neurons, metabolism revs into high gear and burns up the surrounding cells. It takes very little domoic acid to do the same thing. And domoic acid is heat-stable, so cooking the mussels made no difference.

Autopsies were performed on the dead patients with particular attention paid to their brains. The pathologists saw a striking pattern of cell death in the medial temporal lobes, including the hippocampus and the amygdala. Ever since patient H.M. had undergone surgery for epilepsy and lost the ability to form new memories, damage to this area had been known to cause severe memory impairments. The doctors in Montreal called it amnestic shellfish poisoning.

The outbreak represented a major public health problem that had the potential to shut down the entire Atlantic shellfish industry. The source of the domoic acid had to be found quickly, before it killed again.

A team of biologists fanned out over Prince Edward Island. Led by Stephen Bates, a biologist employed by the Canadian Department of Fisheries and Oceans, they collected water samples and mussels from all the inlets—twenty-five separate sites. Mussels feed on phytoplankton, so the biologists focused on skimming the surface water, where the algae accumulated. This task required towing fine-meshed nets through the inlets and pumping water from the estuaries where boats couldn't reach. It was a race against time to beat the seasonal freezing of the bays.

The contaminated mussels were concentrated on the eastern seaboard of the island, centered around the coastal hamlet of Cardigan. Apart from the toxic levels of domoic acid in the mussels, the only other distinguishing characteristic of the water was a high concentration of a phytoplankton called *Pseudo-nitzschia*. Bates took the *Pseudo-nitzschia* samples back to his laboratory and let them grow. Sure enough, when he tested these cultured samples, they were chock full of domoic acid.

The affected patients had received a toxic dose of domoic acid via the mussels that had eaten the plankton. It was never determined what caused the appearance of *Pseudo-nitzschia*, but it disappeared as suddenly as it had arrived. *Pseudo-nitzschia* was a nasty bugger, having caused one of the worst outbreaks of shellfish poisoning in humans ever recorded. The next time it wreaked havoc was in Monterey in 1998, but instead of humans, it was sea lions who suffered.

Frances Gulland was one of the staff veterinarians at the Marine Mammal Center in Sausalito, California. Gulland had been vaguely aware of the Prince Edward Island poisoning, but seeing that it had been a human mortality event, she hadn't paid it too much attention at the time. But in the late spring of 1998, four years after joining the center, she suddenly had cause to remember. That was when Gulland's phone started ringing off the hook.

Most years, the center took in a few hundred seals and sea lions. Usually these were malnourished pups, which meant they showed up in the summer, soon after they were born. The center would take them in, fatten them up, and release them back to the ocean. But the previous winter had been brutal—not for cold weather—but because of El Niño. By May, when the females began to give birth, warm water had been piling up all along the Eastern Pacific for six months. The pups only nursed seven to ten days and then the moms had to head out to sea to catch fish. But they didn't make it very far. One hundred miles to the south, Monterey Bay had been reporting mass strandings of sea lions, and they weren't the usual malnourished pups. These were the moms. And the reports sounded like they were drunk.

It was distressing to see. The animals didn't seem able to walk and were falling over sideways. They moved their heads in odd weaving patterns. They tried to use their flippers to scratch itches that couldn't be satisfied. The worst just collapsed in a quivering mess.

By mid-June, Gulland had taken in seventy adult sea lions, all from Monterey Bay. Fifty-four of them were adult females, and half of those were pregnant. None of the animals were significantly malnourished. Plus, she had to care for the usual round of malnourished yearlings, which had already spiked past three hundred by June.

It was a war zone. Gulland and the staff at the center did their best, using every trick they could think of to help the animals. They started IVs to keep the animals hydrated, since they couldn't eat or drink while having epileptic fits. They pumped the sea lions full of valium and phenobarbital to get the seizures under control. In the worst cases, the animals deteriorated into a state of continuous seizures called *status epilepticus*. It is horrible to see in a human, and it is equally horrible in an animal. *Status epilepticus* only lasts about ten minutes, but the seizures stop only when the brain gets starved of oxygen. At that point, the brain swells. Gulland would infuse high doses of steroids to control the swelling, but it rarely worked, and the animals would soon stop breathing. The pregnant females received ultrasounds, and many of the fetuses were already dead. A dead fetus was a sure route to infection and death, so Gulland had to induce delivery to try to save the mothers.

Nothing in Gulland's training had prepared her for this. By the end of June, she had managed to save just twenty-two of the seventy animals. Of the fifty-four mothers, only twelve made it out alive. Determined to figure out what had killed the sea lions, Gulland set out on a forensic investigation that would consume her spare time for the next two years.

The seizures pointed to some type of brain event. When Gulland examined the brains after death, she found a striking pattern of tissue necrosis. The brains had little holes like Swiss cheese in the medial temporal lobe, right around the hippocampus. At first, Gulland suspected some kind of chemical exposure. But when she examined the cerebrospinal fluid of the dead animals, she couldn't find any trace of the usual suspects, such as lead or mercury.

So it was time to look for unusual suspects. Fortunately, Monterey Bay was one of the most heavily monitored marine sanctuaries in the United States: both the US Coast Guard and the Monterey Bay Aquarium had routinely collected water samples for years. Maybe, she thought, the smoking gun was hiding in those samples.

Every year, before the official start of spring, the sun's rays hit the bay at a high enough angle to stimulate the growth of the microscopic organisms called plankton. The first plankton bloom usually occurs at the beginning of April. Plankton are a major food source for many fish and whales, and any changes to the plankton would have a ripple effect. When El Niño struck, unpredictable things happened.

The average temperature in Monterey in April is normally 54°F. In 1998, it was 62°F. And April is usually a dry month, averaging less than a tenth of an inch of rain. But in 1998, it had rained almost every day for the first two weeks, and by the end of the month, three inches had fallen—thirty times the average.

More than the temperature, the rain was wreaking havoc with the plankton. Gulland thought that the rain had increased the agricultural runoff into the bay. Rich with fertilizer, the runoff provided massive amounts of nitrogen to fuel the plankton bloom. By the end of April, the Coast Guard's water samples showed a new species of plankton.

Pseudo-nitzschia had returned.

When Gulland saw the reports of the *Pseudo-nitzschia* bloom in Monterey Bay, she still had to figure out what that had to do with the sea lions. Sea lions don't eat plankton. They don't have the manual dexterity to eat shellfish like sea otters do, which ruled out mussels as the source as in the Prince Edward Island outbreak. Sea lions eat fish, and they are not terribly picky about it. Gulland started with the most common fish at the scene of the crime—anchovies and sardines. Sure enough, their bellies were chock full of *Pseudo-nitzschia*, which, under the gaze of an electron microscope, looked like shards of glass. Weirdly, the domoic acid in the plankton was confined to the fish guts. Not that it mattered to the sea lions. They ate the fish whole, taking a toxic dose of domoic acid in with their usual food.

After Gulland put the pieces together, her findings were a bombshell. The only difference between the Prince Edward Island event and the Monterey Bay event was the vector for domoic acid. Although El Niño was the catalyst, in all likelihood it was the agricultural runoff that caused the toxic bloom. Nobody had previously linked domoic acid to a major mortality event in marine mammals. In one sense, it was lucky that Gulland discovered it in the California sea lion. The population was healthy and numbered around 175,000, and this was but a blip. But it was impossible to know where *Pseudo-nitzschia* would strike again. If it showed up in Hawaii, it might wipe out the struggling population of monk seals, which numbered a scant 1,400. Gulland's report put the international community on alert to monitor coastal waters—both for the sake of the marine mammals and for the sake of humans.

As the years passed, it seemed that *Pseudo-nitzschia* had disappeared from California, and Gulland hoped it was all behind her. It wasn't.

There was a reason Monterey Bay had been ground zero for the sea lion deaths. The picturesque coastline is the culmination of a magnificent underwater canyon with walls plunging over a mile deep. Intrepid scuba divers can swim out from the shore and look straight down the abyss. The sea lions know this, too. They don't have to swim far to catch fish in the canyon. But the bay is also a funnel for the runoff from the surrounding hills.

With such an abundance of marine life, Monterey Bay was an obvious location to establish a marine sciences laboratory. In 1965, when the University of California built a campus at Santa Cruz, on the northern end of the bay, a marine sciences laboratory was already part of the plan. With a land gift in 1972, construction finally got underway, and the Institute for Marine Sciences became operational in 1978.

The institute's mission was concerned with all aspects of marine life, but the division dedicated to sea lions and their phylogenetic cousins, pinnipeds, was established by Ron Schusterman in 1985. Schusterman had disproven the previously accepted theory that sea lions were able to echolocate like dolphins, and with the new facility, he set out an ambitious research program to probe what sea lions could do, including any ability they might possess to understand language. At the time, apes were the only nonhuman animals thought to have rudimentary language ability. But Schusterman would ultimately show that sea lions could understand sign language. Colleen Reichmuth, a former student of Schusterman's, expanded the research program and eventually took over operation of the pinniped lab when Schusterman retired in 2003. She has been running it ever since.

Reichmuth radiates a warm charm and hugs people both in greeting and farewell. At meetings of marine scientists, she is in high demand, and she can usually be found at the center of a circle of eager students. But her easygoing persona belies the seriousness with which she runs the pinniped lab. For Reichmuth, everything revolves around the animals. At any given time, the lab might house sea lions, otters, and several species of seals. Each animal has his or her own schedule of care, which includes feeding and routine veterinary care. Highly intelligent, they all require frequent mental stimulation. It is a 24/7 facility, staffed by a team of animal behaviorists, veterinarians, researchers, and students of all levels. Nobody joins the pinniped lab and simply starts working with the animals. You cut fish and gradually work up the food chain, learning how to interact with and train the animals.

Reichmuth didn't normally take students in the lab without meeting them. She knew how important it was to filter out the students who just love animals—especially marine mammals. Their sleek forms and big eyes have drawn in countless naïve students. Biologists call it neoteny—the retention of juvenile traits in the adult—and it can be irresistible, especially for those with strong maternal instincts. But neoteny's allure is poison to good science, so Reichmuth had erected strong defenses against touchy-feely students who just wanted to anthropomorphize the animals.

P eter Cook didn't fit the mold of the usual student who joined the pinniped lab. His undergraduate degree was in philosophy, not the usual biology or psychology. In his defense, he had always been interested in the question of animal consciousness, but that would not be enough to clear Reichmuth's standards. To bolster his credentials, he spent a year in Columbia University's post-baccalaureate program in psychology. In his spare time, he also worked at the New York Aquarium in Coney Island, observing the walruses. This was the closest he could get to experiments with wild animals. But, of course, the walruses weren't really wild anymore.

As his time at Columbia drew to a close, Peter applied and was admitted to the PhD program in psychology at UC Santa Cruz. Except for the possibility of another mushy-headed student, this was a no-risk scenario for Reichmuth. The pinniped lab was in the Institute for Marine Sciences. Peter was going to be Psychology's problem. If he worked out with the pinnipeds, great. If not, well, he could always find something else to do in psychology. So, with an uncertain commitment, Peter and his wife, Lilas, moved to the West Coast in 2007.

Life in Santa Cruz suited Peter and Lilas. Peter could bike between campus and the marine lab, and the downtown area had an ample supply of vegan restaurants. The New Leaf Community Market supplied ethically farmed food.

Peter's affable nature made it easy for him to fit in at the pinniped lab. Like everyone else, he started at the bottom and prepared the animals' meals. He picked up the basics of animal training quickly, despite never having any previous experience, not even with dogs.

Peter's first two years in Santa Cruz flew by. Between taking the required courses for a psychology degree, he spent all of his time at the pinniped lab working his way up through the ranks of animal care and training. Learning how to train sea lions was critical for Peter's work. Otherwise, he'd have no way to get the animals to perform tasks that he could use to test some aspect of mental function.

For his PhD dissertation, Peter's idea was to test sea lions' memory at the pinniped lab. Ever since the 1998 mortality event, domoic acid had been identified as the toxic agent responsible for the sea lions' brain damage. The most severe cases had been obvious because of the seizures. But what about the less severe cases? Could damage short of causing seizures affect an animal in other ways? A sea lion with only mild damage might not have seizures but still die from malnutrition because of the inability to find food. A sea lion might even just get lost at sea and drown. If Peter's idea was correct, animals with domoic acid exposure should be impaired on tasks that required them to remember the locations of things, even if they didn't have full-blown seizures. Low-level domoic acid exposure might have been happening even without El Niño.

Answering Peter's question would require a heroic undertaking. Gulland, who was now the senior veterinarian at the Marine Mammal Center, would let Peter know when they took in a sick sea lion. As soon as the animal was healthy enough, Peter would pick it up and bring it to the pinniped lab at Santa Cruz. Under Reichmuth's supervision, he would then train these wild animals to complete a battery of memory tests. To nail down the specific effects of domoic acid, he would also need a control group—sea lions who landed at the center for reasons other than domoic acid toxicity.

Domoic acid doesn't hang around the body. It is excreted in the urine and feces within forty-eight hours. So unless an animal had just eaten a bunch of toxic anchovies, it would be difficult to tell whether it was suffering from domoic acid toxicity or from some other problem. The severe cases would have the telltale neurologic symptoms, but the mild cases would not. It would take an MRI of the brain to see if there was damage to the hippocampus.

Nobody did MRIs of sea lions, at least not as a matter of routine. But they were going to be crucial to figure out which ones had domoic acid toxicity and which did not. Any modern MRI facility should be able to do a basic MRI of a human brain, but sea lions would need to be anesthetized for the procedure. Marine mammals have enormous stores of fat, which could suck up some anesthetics like a sponge. Sea lions also have respiratory systems adapted to long breath-holds, so it required special techniques to keep them ventilated while under anesthesia. Marine mammal veterinarians were rare enough, but those knowledgeable in both anesthesia and radiology were even rarer.

Fortunately for Peter, Sophie Dennison arrived at the center right at the launch of his sea lion project. Dennison was in the middle of her radiology residency when she rotated through the center. Dennison arranged to have the sea lions get an MRI at a facility in Redwood City, which was about halfway between Santa Cruz and the Marine Mammal Center. And luckily for a cash-strapped graduate student, because his project was so unusual, the imaging facility agreed to let them scan the sea lions after hours for free.

P eter got the first call in April 2009. A young male sea lion, named Jettyhorn, had landed at the center for malnutrition.

Peter headed north from Santa Cruz to pick up the sea lion from the Marine Mammal Center. In a couple of hours, he was over the Golden Gate Bridge and turning left to the Marin Headlands to pick up his first subject.

Once there, Peter shooed Jettyhorn into a large dog crate. Jettyhorn was still young, so it was no trouble to lift the crate into the back of Peter's truck. Peter didn't linger, because he knew the trip would be stressful for Jettyhorn, who would be riding shotgun in the back of the pickup. To minimize the stress of transport, Peter avoided the interstate and took the scenic route along the Pacific Coast Highway. Around Half Moon Bay, at the famous surfing spot everyone called Mavericks, Jettyhorn perked up to the sound of the waves. Peter could see him waggling back and forth in his crate, craning his neck to see the ocean. "Sorry, not this time," Peter said.

Because he was a wild animal, Jettyhorn was put into the

lab's quarantine pool, which Peter had converted into a giant maze. The plan was to give the sea lions two tests. Both would require the animal to navigate a simple maze to get a piece of fish. Peter knew he would have a limited time to work with each animal, so he needed a test that could be trained quickly, even in a wild sea lion unaccustomed to working in captivity.

The first test was designed to simulate how sea lions find food. Sea lions are foragers and naturally curious, which made them ideal subjects for maze-tests. If they knew there was a fish to be had, they would explore. Once a day, Peter placed four buckets around the perimeter of the enclosure that surrounded the quarantine pool. Only one of the buckets contained a fish, and this bucket would be in the same location every day. So that the animal wouldn't see him put the fish in the bucket, Peter rigged up all the buckets with ropes and pulleys. Every day, he would load the same bucket, and then lower all of them into the pen at the same time.

On the first day, Peter lowered the buckets and Jettyhorn hopped out of the pool to investigate. Maybe it was bad luck, but he first went to the bucket farthest away from the one containing the fish, and then made his way counterclockwise around the perimeter, checking out each bucket until he got to the correct one. Peter timed it at thirty-two seconds.

The next day, Jettyhorn hopped out of the pool and seemed to think about it. He just stood on the deck, waggling back and forth. After twenty-eight seconds, he wriggled over to the correct bucket, without visiting the others. By the third day, he was even better, getting his fish in twelve seconds, and by day five, he had mastered the task, making a beeline to the bucket in three seconds.

Jettyhorn continued doing well with his daily foraging test, so Peter introduced a more complex test of memory. He based it on a common protocol used in laboratory rats, called delayed alternation. In this test, a rat is placed at the long end of a T-shaped maze. In one of the arms is a piece of food, and all the rat has to do is find it. If the rat goes to the wrong arm, he is blocked from going to the correct one, and the trial ends. On successive trials the food is placed on opposite ends of the T, so if the rat wants to eat, he must remember where he just went.

To make a sea lion version of the delayed alternation test, Peter built a giant maze in the quarantine pen with the pool at the base of the T. In the first phase of the test, the sea lion would have to jump out of the pool into the maze, waddle down the chute, and then exit either to the left or to the right. So that the sea lions couldn't solve the puzzle by scenting, Peter hid behind a blind outside of the pen. If the animal exited the correct way, he tossed a fish into the enclosure.

This was pure operant conditioning. Jettyhorn could not know what was expected of him, but his natural curiosity led him to explore the maze. If he exited the correct side, he got fish. But he also had to make his way back to the pool to reenter it. Returning through the way he had just come would not do.

It took a lot of patience. Peter and Reichmuth had previously decided that an animal would need to get 85 percent of the trials correct before moving on to the next phase. Day after day, Peter would come at feeding time to make Jettyhorn work for his fish.

Finally, after a couple of weeks and 421 repetitions, Jettyhorn seemed to know what was going on. It was time to make it harder by taxing his memory.

During the training phase, the only delay was how long it took Jettyhorn to exit one side, eat his fish, and return to the other end of the maze, which was not a long enough delay to tax the hippocampus. Now, Peter would block the entrance to the maze for seven seconds. While he waited, Jettyhorn would have to remember which side of the maze he had just exited. They would do forty trials, and Peter would tally up the number of correct choices.

Testing day was a big event. Peter had spent two years planning this experiment, and after several weeks training Jettyhorn, Peter was nervous about how he would perform. He didn't expect perfection, but he had grown fond of Jettyhorn and hoped that he would do well.

Jettyhorn didn't disappoint. Unaware that it was anything other than a usual day, he got thirty-seven out of forty delay trials correct—well over 90 percent, and certainly not by luck. Reichmuth was pleased, too. Peter's audacious idea was going to work.

It was a bittersweet success. With his task completed, Peter had to take Jettyhorn back to the Marine Mammal Center. Jettyhorn appeared healthy, and his performance on the delayed-alternation test showed that his memory was intact, so he would probably be released back into the ocean. A happy outcome, but Peter would still miss him.

Spring and summer brought a trickle of patients to the center, and ultimately to Peter's study. It was fortunate for the sea lions that El Niño didn't reappear, but the favorable weather patterns slowed Peter's experiment. Months would go by without a subject for his study. In the end, it took nearly six years to get to his target of forty animals, but Peter finally showed the link between domoic acid exposure and memory function in the sea lion.

The MRIs provided the missing link.

When Peter compared the size of the hippocampus with the sea lions' performance on the delayed-alternation test, he found that the smaller the hippocampus, the more errors a sea lion made. Nearly twenty years after Gulland's initial discovery, the team had a plausible mechanism for the sea lions' strandings. In years with toxic algal blooms, the anchovies ate the algae, accumulating domoic acid in their guts. The sea lions ate the anchovies, each time taking a small hit to the hippocampus. Over time, these exposures accumulated to the point of impairing an animal's ability to forage for food and navigate. Eventually the affected sea lions died or stranded on the beach in a malnourished state. It was a tour de force study following an ecological toxin all the way up the food chain to the brain and its effects on cognition.

Although Peter's team was circumspect in pointing fingers at the cause, it seemed clear that these "unusual mortality events" could not be blamed solely on El Niño or other natural phenomena. Excessive rain was the catalyst, but agricultural runoff was the proximal cause of algal blooms. The phytoplankton bloomed because of the increase in nitrogen in the water, which was caused by human agricultural activity.

Nobody knows if the Prince Edward Island event was also due to human activity, but the effect of domoic acid in the patients from Nova Scotia shows what it's like for a sea lion to experience the same disorder of consciousness. And because seizures are common in people, we know exactly what it feels like. In a generalized seizure, the entire brain is affected, while in a partial seizure only a subset is involved.

The most common site for partial seizures is the temporal lobe, right around the hippocampus, and this is the same region affected in the sea lions. The nineteenth-century Russian novelist Fyodor Dostoyevsky was thought to suffer from temporal lobe epilepsy, and he wrote vivid accounts of his fits. In *The Idiot*, Prince Myshkin, a character based largely on Dostoyevsky's own experiences, described it like this:

During his epileptic fits, or rather immediately preceding them, he had always experienced a moment or two when his whole heart, and mind, and body seemed to wake up to vigour and light. In the last conscious moment preceding the attack, he could say to himself, with full understanding of his words: "I would give my whole life for this one instant."

This lasted perhaps half a second, yet he distinctly remembered hearing the beginning of the wail, the strange, dreadful wail, which burst from his lips of its own accord, and which no effort of will on his part could suppress.

Next moment he was absolutely unconscious; black darkness blotted out everything.

The sufferer was immediately taken to his room, and though he partially regained consciousness, he lay long in a semi-dazed condition.

A sea lion would not have the sentimentality of Prince Myshkin, but given the similarity in brain anatomy, it is very likely that the subjective experience of a temporal lobe seizure is very much the same, as is the confusion afterward. But when it happened to a sea lion at sea, the animal probably drowned. Peter's work had drawn the link between domoic acid exposure and damage to a sea lion's hippocampus, but there was still much that was unknown. The worst of the sea lions had obvious shrinkage of the hippocampus, but was that the whole story? The hippocampus was connected to many parts of the brain. How did hippocampal damage affect these connections? It was likely that disrupted communication might be one of the earliest effects of domoic acid toxicity, even before there was gross damage to the hippocampus.

Although Peter had collected MRIs on the sea lions, these questions couldn't be answered by simple structural scans. They were questions about connectivity. For that, a different type of scan was needed, and that is how I became involved.

Peter had already been working in my lab at Emory for a year when he arranged for the Marine Mammal Center to send us the brains of sea lions who couldn't be saved. The first one showed up double-bagged in plastic with a few tablespoons of brownish liquid to keep it moist. It was about the size of a slightly flattened grapefruit. Someone had scribbled the sea lion's name on the bag: *Little Hoot*.

I tried to imagine what Little Hoot had been like. Diminutive. Unbearably cute. Probably a bit of a clown.

It made me sad to hold his brain. Certainly I had seen plenty of brains before, both human and nonhuman, and oftentimes I had known the deceased owners' names. Human names didn't say much about someone's personality, but animal names did. Or at least we thought they did.

"It's not very big," I said.

"He was probably a pup," Peter replied.

There was no point in grieving over Little Hoot's death.

I didn't even know him. The best I could hope for was to lend some meaning to it by learning something about sea lion brains. Peter and I thought we might be able to map out the hippocampal connections by using a technique that had become common in the human neuroimaging world. It was called diffusion tensor imaging, or DTI.

DTI, an MRI-based form of imaging, takes advantage of the fact that the movement of water molecules around the brain is biased. If the brain was a uniform mass of tissue, a water molecule would bounce around randomly with no direction being more likely than another. But brains are highly structured. The white matter, which contains the axons of the neurons, form the roads and highways of the brain. A water molecule deposited in the white matter would have a hard time moving across the fibers, because white matter is made up of mostly fat and cholesterol—substances that do not mix with water. Instead, a water molecule takes the path of least resistance and moves in the same direction that the axons are running. By measuring the preferred direction of water movement at every location in the brain, DTI creates a map of the white matter.

In order to measure the effects of hippocampal damage on the rest of the brain, we would first need to map out all the white matter connections. Nobody had ever done this in a sea lion brain, so we had no guide of where to begin.

Peter's plan was to obtain as many brains as possible, some with domoic acid toxicity and some without. Then, we could see how the domoic acid disrupted the normal patterns of connectivity. Doing this type of mapping would be tricky in a dead brain. Without blood coursing through the vessels, the brain would not give off nearly as much signal in the scanner as a live one. On the other hand, we wouldn't have to worry about the subject moving.

Although it was a bit of a niche area of research, a few groups had successfully done DTI in postmortem human brains. It took a long time—upward of twelve hours—and the signals were weak, which limited the quality of the reconstruction of the white matter tracts. Karla Miller, an MRI physicist at the University of Oxford, had been trying to improve the quality of scans by experimenting with alternative ways to program the MRI software. By 2012, she had achieved some success, showing that with her approach she could obtain high-quality DTI data in postmortem brains in half the time that conventional sequences took. Strangely, her results had not attracted much attention, probably because not many people were interested in scanning dead brains. But I found her results fascinating and reached out to her to see if she would like to help with the sea lion project. She was delighted that someone else was interested in using her sequences and promptly sent us instructions for installing them on our scanner.

In preparation for scanning, Peter set Little Hoot's brain in agar inside of a Tupperware container. To prevent the water in the gelatin from swamping the signal in the brain, he first doped the agar with a small amount of gadolinium. Gadolinium is a rare-earth metal that has very desirable magnetic properties. It is most commonly used as an intravenous contrast agent for clinical MRIs. It is particularly useful in revealing where tumors are located. In our scans, though, gadolinium would attenuate the signal coming from the agar and help the brain's signal stand out from the background.

We needed to do approximately eighty different scans. Some of these were souped-up structural scans to obtain detailed anatomy, but the DTI scans formed the heart of the protocol. For each of these scans, a magnetic field would be applied

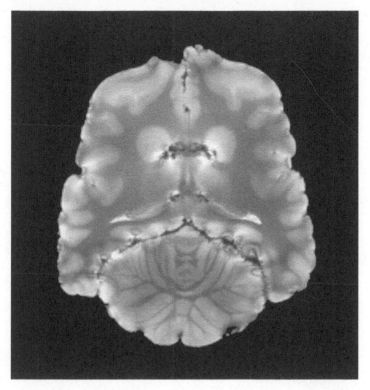

Little Hoot's brain. (*Gregory Berns*)

in a specific direction. The scanner would then measure, at every location in the brain, how water diffused in that direction. This would be done fifty-two times, each in a different direction. The scanning itself would take about six hours. Later, we would reassemble all the scans and calculate the preferred diffusion directions in the white matter.

The structural scans looked great and everything seemed to be ticking along smoothly. With the preliminaries out of the way, it was time to cue up the main event—the DTI scans. We loaded everything according to Karla's protocol and hit *scan*.

The scanner clicked a few times and launched into the sequence, which was heralded by a loud buzzing. And then it stopped after only a second. A red line appeared on the scanner console, indicating some type of failure.

"What happened?" Peter said.

"It looks like the gradient power was exceeded," I replied.

The gradient coils consumed a lot of power, and this heat had to go somewhere. If the heat could not dissipate, the copper would ignite the surrounding material. To prevent the outbreak of a fire, the gradients contained cooling pipes, but even with the cooling system, the coils would heat up. If they got too hot, internal temperature sensors would trigger an automatic shutdown. At least that was what was supposed to happen. The scanner software also had several checks and balances to prevent an unwitting operator from asking it to do something dangerous. Which apparently was what we had just done.

DTI demanded a lot from the gradient coils. The stronger the magnetic field that was applied, the better we could detect the water diffusion. But the scanner software had just told us that it could not perform the requested scan safely. If we attempted to override the warning, there was a real chance that we would damage the magnet.

"We'll have to back off the gradient," I said.

"By how much?"

I shrugged. "Until the scanner says it's safe."

Once we hit the safe point, we settled back and let the scanner run for the next six hours. By early evening, the scan was complete, and without any damage to the gradient coils. It would take another hour to upload all of the data to the laboratory computer, so we set that in motion and went home for the night.

The critical part of the analysis took all fifty-two diffusion images and calculated the distance that water molecules diffused and their preferred direction. This had to be done at every location in the brain, which added up to 350,000 voxels. Karla had written software to make these calculations but had also warned us that unless we had a supercomputer to run it on, it could take a while. We didn't have a supercomputer, but the main computer in the lab had a decent amount of memory (72 GB) and sixteen CPUs. By running the CPUs in parallel, each could work on a different location in the brain, and the calculations could be made simultaneously. Even so, it took two days to process Little Hoot's brain.

To visualize the DTI results, we used different colors to denote the principal directions that the fibers ran in: red was for fibers running left-right, green was for forward-backward (more precisely, rostral-caudal), and blue was for top-bottom (or dorsal-ventral). The images were gorgeous, far better than either Peter or I had expected. A swath of red fibers dominated the center of the brain, clearly showing the corpus callosum, which connected the left and right hemispheres. Laterally, we could see tracts running between the front and the back of the brain, and these were interwoven with fibers running between the top and bottom.

By the time our DTI studies got underway in 2015, El Niño had, once again, returned. It dwarfed even the 1998 El Niño. This time, sea lions were stranding all the way south to San Diego, and the Marine Mammal Center kept sending us the brains of the animals who had to be euthanized. Many, but not all, had had seizures. By the end of the summer, we had eight brains. We set out to see what damage the domoic acid had done.

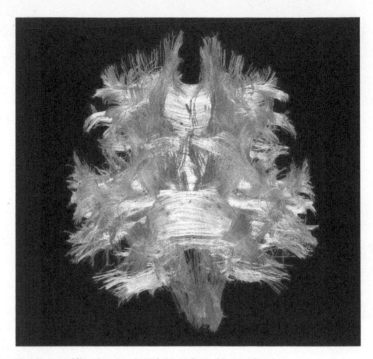

Visualization of fiber directions in Little Hoot's brain. (*Gregory Berns*)

In temporal lobe epilepsy, the hippocampus eventually becomes scarred and shrinks. But the brain tries to compensate. As the hippocampus gets increasingly damaged, the remaining neurons have to perform more and more work. Paradoxically, the downstream effect is an increase in connectivity. If domoic acid caused brain damage primarily through its promotion of seizures, then we would expect to see the same paradoxical effects on the hippocampus as had been observed in human epilepsy.

Although the 3D reconstructions of the sea lion's white matter tracts were beautiful, we needed a more precise way of measuring where the hippocampal pathways were going. In human neuroimaging, it had become common to use a technique called probabilistic tract tracing. The idea was to place a digital seed at a location in the brain. Using the predominant directions in which water diffused, you could simulate the movement of a hypothetical water molecule from the seed location. Then, at the new location, you did it again, and so on. It was hopscotching through white matter. Because the preferred directions were never absolute, there was always some uncertainty about which way the hypothetical molecule would move. It was standard practice to run these simulations thousands of times and average the resultant pathways. Peter placed digital seeds within the hippocampus and instructed the software to reconstruct the tracts emanating from it.

The answer was immediate. The number of tracts to the thalamus was greater in the sea lions who had domoic acid exposure—just like in humans with epilepsy. And the similarity in brain pathology between sea lions and humans pointed toward a similarity of experience. So we now knew what it was like for a sea lion to experience domoic acid poisoning. It was the same as for a human. But did analogous experiences run deeper than that?

The answer lay in the brain of an unusual sea lion—one with a knack for rhythm.

Chapter 5

Rudiments

Ronan was born somewhere along the coast of northern California in 2008. By the age of one, she was an incorrigible wharf rat. Probably weaned too early from her mother, Ronan would hang around docks pestering people for handouts. Sometimes she waddled down Highway 1, amusing tourists but disrupting traffic. Whether she got lost and couldn't find her way back to the ocean or was simply looking for her mother, nobody could say, but her frequent walkabouts on a busy road eventually landed her at the Marine Mammal Center.

There didn't seem to be anything wrong with Ronan, and after the usual fattening-up treatment, she was returned to the ocean.

And then she would show up again on some other stretch of the Pacific Coast Highway.

After Ronan bounced back to the center for the third time, her fate was uncertain. She didn't have much experience foraging for her own food, and her rescuers didn't think she could survive in the wild. Unlike many of the other sea lions, she was relatively healthy. After a week or so of plumping her up, there would be no medical reason to keep Ronan at the center.

It was a tough call for senior veterinarian Frances Gulland. At any given time the center might house sea lions, harbor seals, fur seals, and even the giant elephant seals, whose main breeding ground was halfway down the coast at Año Nuevo. The center had limited space, and every year seemed to bring more and more animals.

Everyone at the center liked Ronan, and it was her affinity for people that landed her there as a repeat offender. But the center wasn't meant to amuse animals or humans. If they couldn't find a home for Ronan, she might end up at a marine park. Or being euthanized if nobody wanted her.

Peter Cook had needed control subjects for the domoic acid study. These, in fact, were hard to find. Healthy sea lions didn't normally strand themselves. Mostly it was the sick ones who did. So Peter was delighted when Gulland called him to pick up Ronan. Working with Peter and Colleen Reichmuth at the pinniped lab would at least buy time for Ronan before they had to decide her fate.

Ronan was a quick study and had no trouble with Peter's alternation task. Some of the sea lions with domoic acid toxicity could take months to train, but Ronan breezed through the whole thing in a month. Her contribution to science completed, she would soon need a home.

Sea lions could live for thirty years in captivity, so adopting one was a long-term commitment—one that could extend beyond Reichmuth's career at the pinniped lab. Besides, one of the other subjects, a sea lion named G-Dock, had become particularly popular. G-Dock had taken to toddling around her pen with one of Peter's buckets on her head. She had a goofy, fun-loving personality that endeared her to the other trainers, and Reichmuth was thinking about keeping her. But Ronan did better on Peter's tests. She ultimately beat out the class clown for permanent residency status, and another facility adopted G-Dock.

It was lucky for everyone that Ronan stayed. She would soon prove to have unique skills that would change how people thought about the evolution of language.

If you ask people what differentiates humans from other animals, most will say it is language. Indeed, no matter how much we anthropomorphize the behaviors and vocalizations of different animals, there is no denying that only humans have full-fledged language ability. Many have theorized that the so-called "language instinct" was the fundamental evolutionary innovation that gave rise to hominids. And because other animals do not use words to label experiences, some scientists have questioned whether we could ever know what it's like to be another animal.

But, as I was to learn from Ronan, language is just the tip of the iceberg. Beneath the surface lies a wealth of cognitive processes that humans do, in fact, share with other animals.

For humans, a word is laden with meaning. Even a simple word, like "flower," conjures up a rich set of images. Children learn early on that words are shorthand for communicating ideas, and that words can represent a universe of animals, things, and actions. At its core, language is about the abstract representation of concepts. A word is but a stand-in for the thing it represents. Even though other animals don't speak, many can learn vocal commands. The million-dollar question is whether they are simply responding in a reflexive manner to sounds, or they understand that spoken words have meaning. It boils down to symbolic representation in the brain.

Words come in different forms—spoken, written, signed—some of which may be easier for animals to process than others. In the 1980s, Ron Schusterman began the investigation into sea lions' symbolic capacity by using hand signals. Schusterman used arbitrary gestures to teach sea lions a large vocabulary for objects (such as bats, balls, and rings), modifiers (big, small, white), and actions (fetch, tail touch, flipper touch). His favorite sea lion, Rocky, could respond to over 7,000 combinations of gestures. Since it was unlikely that she had memorized each unique combination, Schusterman concluded that Rocky had, at some level, internalized the meaning of individual gestures and was able to understand simple phrases.

Reichmuth thought that sea lions like Rocky, and later, another sea lion named Rio, were doing something much more sophisticated than simply associating sounds with actions, but not to the point where words had become symbols. Instead of using hand signals, she tested sea lions' capacity for understanding pictorial symbols. Schusterman had shown their affinity for hand signals, but pictograms were decidedly more abstract. This abstraction made the pictograms particularly well-suited for studying the logical operations that might underlie language comprehension. The pictograms were made from square pieces of plywood, one foot per side, with the figures drawn in black paint on a white background. Like a library of vinyl records, the pictograms still line the shelves in the pinniped lab.

Reichmuth demonstrated that the sea lions could learn simple *if . . . then* rules. In one experiment, Rocky learned that whenever she saw a spiral pictogram, she should nose another pictogram of a rectangle, even with distractor pictograms present. Importantly, this was an arbitrary association that Reichmuth specified. When new pictograms were introduced, Rocky was able to learn the associations more quickly by ignoring pictograms that she already knew were associated with something. It's called learning-by-exclusion, which is a rather sophisticated cognitive process that requires the animal to understand the overall context of the task and to remember which things have already been incorporated into rules. More impressive than that, Reichmuth showed that Rio could link together logical relations. When taught *if spiral . . . then rectangle* and *if rectangle . . . then circle*, Rio would correctly respond to *if spiral . . . then circle*. This logical operation is called transitivity and is foundational to human language. Rio could even reverse the direction of the logic, correctly responding to *if rectangle . . . then spiral*. This operation is called logical symmetry, and it allows humans to recognize the equivalence of statements like: *Sally hit the ball* and *The ball was hit by Sally*.

Rocky and Rio had demonstrated their capabilities for logical operations and the fact that they solved logic problems in much the same way humans do. But while logic is necessary for language, it isn't the whole story. Language, especially when spoken, has a rhythm. Take the natural rhythms of speech, exaggerate them, and you have the rudiments of music. If Rocky and Rio could do some of the logical operations underlying language, maybe sea lions also had some rudimentary musical ability. And this is where Ronan's and Peter's talents shined.

Charles Darwin was intensely interested in music and thought that, like all things in humans, it must have its origin in other species. In *The Descent of Man*, he wrote, "The perception, if not the enjoyment, of musical cadences and of rhythm is probably common to all animals, and no doubt depends on the common physiological nature of their nervous systems." Not everyone agreed with Darwin's theory on the origin of music. His contemporary, Herbert Spencer, wrote an essay in 1857 titled "The Origin and Function of Music," arguing that speech was the necessary predecessor to music, and therefore, solely the province of humans. Not much changed for a long time after the Darwin vs. Spencer debate. To find support for Darwin's theory, one would need to find evidence of musical ability in animals.

It is unlikely anyone will ever find a sea lion that can sing. Even the best vocal mimics in the animal world can't carry a tune. But singing is a high bar. Peter realized that there were many aspects of music besides melody that one might potentially find in animals. The most basic is rhythm. If it doesn't have a beat, it isn't music, although the best music has more than a beat. Good music grooves. All of the great guitarists—Jimi Hendrix, Carlos Santana, Keith Richards, Stevie Ray Vaughan— have been masters of the groove and can create a pocket that exerts an irresistible pull on a listener. Just listening to music with a good groove makes us move in time.

Locking into a rhythm is called synchronization, or sometimes, entrainment. Seeing whether animals could do it provided a way to put Darwin and Spencer's debate to rest. Surprisingly, no one had tried until 2006, when Aniruddh Patel, a neuroscientist in San Diego, proposed the vocal mimic theory of rhythm. Patel suggested that vocal learning, whether in a human or, say, a parrot, required precise synchronization between auditory inputs and motor outputs. Speech requires both that we perceive the precise timing inherent in speech and that we are able to reproduce that pattern with our vocal chords. Although it sounds complicated, humans can do it without thinking about it. And if it works for producing speech, then it should work for reproducing rhythms. By the principle of parsimony, Patel suggested that the brain mechanisms needed for vocal mimicry and beat synchronization were one and the same.

Patel's vocal mimicry hypothesis led to a strong prediction: only animals that are vocal mimics should be able to entrain a rhythm. It was an interesting idea, if something of a redundant point. After all, nobody had documented an animal moving in time to a rhythm.

The idea went nowhere until a few years later, when, out of the blue, a singular bird became a viral sensation on YouTube. Snowball was a sulphur-crested cockatoo who made the rounds of talk shows because of his head-banging, claw-stomping dance moves in time to a bunch of pop songs. His favorite was the Backstreet Boys' "Everybody."

It was hard to tell from the videos exactly what Snowball was doing. Was he just generally bobbing up and down, or was he really in sync with the beat? So Patel analyzed the videos frame by frame and found that Snowball really was staying in time. Of course it helped that the song had a strong beat.

That was when Peter got interested in sea lions and music. According to Patel's vocal mimic theory, a sea lion should not be able to synchronize to a beat. And although sea lions can grunt and bark, no sea lion had ever shown anything close to vocal mimicry. In the downtime between subjects for his memory study, Peter thought he could teach Ronan to bob in time to a beat, and if they were successful, that would pretty much be the end of the vocal mimic theory.

Animal training works best when you can start with a natural behavior. From there, the trainer can gradually shape the behavior into something else. It's like a treasure hunt, and all the trainer can do is tell the animal "warmer" or "colder." Although sea lions have remarkable fin dexterity, they have even better control over their neck muscles. Peter thought teaching Ronan to bob her head should be a piece of cake. It didn't take long to teach Ronan to follow Peter's hand in an up-and-down motion. Each time she moved her head in the direction his hand was moving, Peter would blow a whistle, indicating to her that she had done the right thing. It was the animal equivalent of "warmer." After each whistle, Peter would toss her a fish.

Most animals have brief attention spans. Frequent, short training sessions are more effective than long ones. Sea lions seem to be an exception, and Ronan seemed to never get tired of training, except when she got frustrated. Working just on weekends, Ronan was happily bobbing to Peter's hand within a few weeks, and she was ready for the introduction of an auditory cue. Peter hoped that Ronan would transfer what she had learned to a metronome click, and he would be able to gradually remove the visual cue of his hand.

While piping an electronic metronome through a loudspeaker, Peter conducted Ronan as his orchestra of one. They began with a medium tempo speed of 120 beats per minute (bpm). It took weeks before Ronan made the connection between the metronome and her head bobbing, and Peter could gradually fade out his hand motions.

Next, Peter had to rule out the possibility that Ronan had simply learned to bob her head 120 times per minute, rather than learning to bop along with the metronome. So Peter introduced a new rate of 80 bpm, alternating sessions at the two tempos to teach Ronan "rate-flexibility."

Ronan appeared confused by the new tempo. Rather than locking in to the beat, she began to bob at different rates within each session. And because she wasn't synchronizing, Peter didn't toss her any fish. Lacking reinforcement, she became frustrated and quickly retreated to her pool.

Seeing that Ronan was confused, Peter simplified the task by training only with 120 bpm or 80 bpm on any given day. And instead of waiting for twenty successful beat synchronizations, as he had before, he rewarded Ronan for just two successful beats. Gradually, over several months, he lengthened the number of successful bobs required for a piece of fish.

After six months of training, Ronan finally managed to entrain to both the 80 and 120 bpm click tracks. To go further in making sure she was really synchronizing, Peter tested her on tempos that she had never heard. Ronan locked in at all but the slowest. This would have been enough to disprove the vocal mimic theory, but Peter wanted to see if Ronan, like Snowball the cockatoo, could synchronize to a musical beat.

Although dancing comes naturally to most humans, it's actually quite a complicated skill. Locking in to a musical beat requires a person's brain to pick out the beat from within a complex polyphonic stimulus. The task is made somewhat easier because music usually follows a repeating pattern of three or four beats. Pop music almost always has four beats per measure, with the bass emphasized on beats one and three and a snare drum on beats two and four (called the backbeat). James Brown hit the one-beat hard ("Get Up Off of That Thing" or "Sex Machine"), while a lot of early rock-n-roll emphasized the backbeat—like the snare in Elvis Presley's "Hound Dog." Jazz and blues often superimpose a long-short alternation between beats, called syncopation, or for the hip, swinging the beat.

To get Ronan ready for her dance debut, Peter began mixing in a backbeat with her primary training. The backbeat was played at a different volume and alternated with the primary beat. Luckily, that didn't bother Ronan at all. She continued bobbing in time.

The ultimate proof of Ronan's beat synchronization would come from "Everybody," just like for Snowball. But Ronan couldn't train on that song. That would be cheating. Ronan needed to be trained on another song and then tested on the Backstreet Boys. This left Peter with a momentous decision. What should be Ronan's first song?

"Needs more cowbell," Christopher Walken declared in his famous *Saturday Night Live* skit.

Indeed. Cowbell seemed to be the ticket for Ronan's musical debut. A cowbell cut through a musical mix and was pretty close to the tones Peter had trained her on. He couldn't lift directly from the SNL skit and use Blue Oyster Cult's "Fear the Reaper," because the song had too much going on musically. Plus, it's just not a dance song. He needed another cowbell song—something with a strong beat. In the end, the winner was Creedence Clearwater Revival's "Down on the Corner." The song opens with four beats of the hi-hat, and then the cowbell comes in, and it stays on every beat through the entire song. Nothing syncopated, nothing fancy, just a strong 4/4 beat the whole way through. Perfect for dancing sea lions.

Too bad Ronan didn't synchronize with CCR's song right away. When the music started, Ronan bobbed for a bit, but she was clearly not in time, so Peter didn't toss her any fish. And although sea lions can be highly motivated, Ronan was of the mindset, "No fish, no work." She went back to her pool.

Peter, though, had become much better at animal training than he'd been when starting out with Ronan, and he knew Ronan's quirks. He knew, for example, that after a little break Ronan would come back. After all, fish were still up for grabs. Sure enough, after about ten minutes, she was back out of the pool ready to work. This time around, he looped a simple subsection of the song without vocals.

After about ten sessions, Ronan had found her groove and was reliably bobbing for twenty consecutive beats. Peter judged her ready for the Backstreet Boys, and Ronan nailed it. She locked into the native tempo of 108 bpm and stayed with it throughout a 90-second clip. Verse, chorus, bridge—it didn't matter. And she wasn't ahead or behind the beat, either.

Critics might argue that Ronan had simply locked into the initial beat. Because "Everybody" never varied in tempo, it might appear that Ronan had stayed with the music but was, in fact, just following her own internal rhythm. To head off this criticism, Peter tested Ronan on "Boogie Wonderland" by Earth, Wind, and Fire. "Boogie Wonderland" does not have a steady tempo—it varies from 123 to 138 bpm. Ronan nailed that, too, a feat which, when it hit the news, prompted one of the band members to remark that the band should be renamed "Earth, Wind, Fire, and Water."

Although it is impossible to scan the brain of a dancing sea lion, there is plenty of data about what happens in the brain of a human doing something rhythmic. When I began using fMRI in the 1990s, finger-tapping experiments were very popular. It was common to map out the motor region of the cortex by having participants tap their fingers together in the scanner. If a person did this with the right hand, activation in the left motor cortex, around the central sulcus, occurred. I had been particularly interested in what happened when a person tapped increasingly complicated rhythms.

To investigate this question, back in the early 2000s some colleagues and I had gotten human participants into the MRI scanner and instructed them to begin tapping a simple, steady rhythm. After twenty consecutive beats, I cued them to increase the complexity of the tapping, so they would have a syncopated rhythm alternating between long and short intervals, like dah-di-dah-di-dah. After that, they had to increase the complexity to a three-interval pattern, like di-dah-daah-di-dah-daah. Although they didn't know it, they were tapping out Morse code. When we examined what was happening in their brains, we observed two things. First, the faster they tapped, the more active the motor region was. More interesting was the second observation: higher complexity was linked to activity in the cerebellum. In mammals, the cerebellum contains 80 percent of the neurons in the brain, and it is thought to be the critical structure for timing and coordination of movement. The cerebellum has also been linked to cognitive processes, and even autism.

In all, Peter had spent over a year training Ronan to keep the beat. What began as a weekend diversion had displaced his main research on the effects of domoic acid and added a year to his time in graduate school. But Peter had no regrets. It was important to show that beat synchronization did not depend on vocal mimicry and was therefore a more general skill found across the animal kingdom. And even after Peter joined the Dog Project, he continued collaborating with Colleen Reichmuth. So when the occasion arose that Peter needed to go back and observe progress on a new rhythm experiment with

Ronan, I asked if I could tag along. It was a good excuse to go to Santa Cruz and meet Ronan.

The seals and sea lions were housed in a line of low-slung pens. It looked like a large dog-run, except that each pen had a pool of ocean water and a connecting deck for sunbathing. But unlike in a dog kennel, nobody was barking.

Peter was nervous. It had been three years since he had last seen Ronan, and even though he called her his surrogate child, he wasn't sure she would remember him.

Peter looked toward the pens and said, "We'll see if she recognizes me."

It was still quiet, and there was no sign that the sea lions were aware of the tension building. "I won't be offended either way," Peter said. "Well, maybe a little."

Peter approached the sea lion pens for his long-awaited reunion. Rio was in the first pen, swimming lazily around her pool. "Hi, Rio!" Peter said.

Rio hopped out of her pool and toddled over to investigate. She gave Peter a friendly sniff and went back about her business.

Ronan was in the next pen. Younger and more active than Rio, Ronan was alternating between waddling around on the deck and diving into her pool.

Peter called out, "Ronan! You look so big!"

It wasn't clear that Ronan heard Peter, and, unlike a dog, she didn't seem to recognize him visually.

Peter turned to Andrew Rouse, who had worked with Peter on the rhythm experiments, and said, "Andrew, could you call Ronan?"

Andrew clapped on the pen as if he had some fish.

Ronan immediately jumped out of her pool and came over to investigate.

Peter put his face up to the pen and blew on Ronan's nose. "Sea lions greet each other this way," Peter said.

Ronan sniffed deeply and snorted. Seemingly satisfied, she turned away and went back to her pool activities. It was impossible to say whether that was a sign of recognition. Certainly Ronan didn't react like a dog greeting his human after a long absence.

Peter looked wounded. "Oh well, what can you do?"

With perfunctory greetings out of the way, Andrew set up the speaker for the new experiment. This time, they were going to see how Ronan adapted to perturbations of a rhythm.

This type of experiment had been done in humans and was accomplished by throwing in a beat either a little early or a little late. The bigger the perturbation, the longer it takes to reestablish the rhythm. The process can be analyzed as if the person's brain and body were acting like a bunch of springs. The springs are meant to represent the oscillatory behavior in networks of neurons. The acoustic stimulus sets the springs vibrating, and at the other end of the chain, the person's finger moves in time. When a beat comes in either early or late, the system goes out of whack and takes a while to start resonating in time again. The dynamics of that process are determined by the stiffness and coupling of the springs, which are dictated by the brain and muscles. By measuring how people adapt to perturbations, one can deduce simple properties about the coupling of auditory and motor systems in the brain.

Andrew started piping the metronomic sound through the speaker. It was a sinusoidal sound, varying up and down in pitch like a European police siren, only slower. Ronan knew what to do. She stood in front of the speaker and started bobbing up and down. The synchrony was uncanny. After twenty or so beats, Andrew tossed her a fish.

After a few rounds of this, Andrew played a recording that began identically and then changed. On this particular trial, the change was heralded by a sudden increase in the tempo. Ronan lost the synchronization, but within a few beats had recovered and was back in time. On the next round, the change occurred with a beat appearing early but without a change in tempo. Called a phase shift, this also temporarily threw Ronan, but, as before, she recovered quickly.

Peter's team would eventually analyze how Ronan reacted to phase and tempo shifts. They would find that she reacted just like humans did. This is remarkable considering how different a sea lion's body is from a human's. Ronan bobs her head, but people tap their fingers (although they sometimes bob their heads too). Regardless of the specific muscles involved, the analysis would show that the neural mechanism coupling a rhythmic auditory stimulus to a rhythmic motor output follows the same dynamics in a sea lion as in a human. Ronan demonstrated that this skill didn't depend on language or an elaborate vocal system. The more likely candidate was the cerebellar system, as I had observed in humans a decade earlier.

Just as Colleen Reichmuth had shown that sea lions have some of the logical building blocks for language, Peter had shown they also have the rudiments for synchronizing auditory and motor systems in rhythm. So even though sea lions don't have full-fledged language systems in their brains, they have at least two components of the brain that contribute to human language. We humans know what it's like to consider *if . . . then* propositions, and even the most rhythmically impaired among

Getting a kiss from Ronan while Peter looks on. (*Colleen Reichmuth* / NMFS 18902)

us know what it's like to dance. Once again, analogous skills and analogous brain networks pointed to analogous experiences.

Watching Ronan dance to the beat, I caught myself bobbing too. It must have been subconscious, but once I realized what I was doing, I knew how she felt. After all, our brains were coupled to the same auditory stimulus. It was like being at a concert with everyone dancing in time to the music. It really wasn't complicated.

After a dozen rounds of dancing and bobbing, Ronan was done for the day. She headed back to her pool, but not before I stole a kiss as she wiggled by. It had been a beautiful day. The sun was sliding toward the Pacific and, against the glare, I could make out two dolphins arcing above the water.

I wondered what their brains were like.

Chapter 6

Painting with Sound

My longtime colleague at Emory, Lori Marino, knows all about dolphins. For her PhD research, she compared the skull anatomy of whales and primates and came to the surprising conclusion that the toothed whales, which included dolphins, had one of the highest encephalization quotients in the animal world.

Through much of the 2000s, Lori continued her study of the dolphin brain. She plugged into the stranding network, and whenever a dolphin died on an Atlantic beach, she jockeyed for access to its brain. Through sheer tenacity, she began to assemble a collection of specimens. She also started scanning them with MRI. Ultimately Lori helped develop much of the modern understanding of dolphin brain anatomy.

Beyond her anatomical work, Lori attracted attention when she teamed up with Diana Reiss, a psychology professor at the City University of New York, to test whether dolphins could recognize their reflections in a mirror. Gordon Gallup, a psychologist, had developed the mirror-test in the 1960s to investigate self-awareness in chimpanzees. The usual way it was administered was to place a chalk mark on a chimp's forehead and observe the chimp's reaction when he looked into a mirror. Touching the chalk is taken as evidence of self-awareness. Humans older than about eighteen months pass the test, as did Gallup's chimps.

Lori and Reiss tested two captive dolphins and found that if the dolphins had marks on different parts of their bodies, including the sides of their heads, they spent more time in front of a mirror than if they did not have the marks. This discovery had a great effect on Lori. If dolphins were self-aware, did they have cognitive worlds like humans? If so, then how could we justify keeping them in captivity?

When I went to ask Lori about scanning dolphin brains, she immediately saw the potential of DTI to answer long-standing questions about their mental lives. Dolphins produce a tremendous range of sounds. Some are thought to be for communication, while others are used for underwater echolocation. Despite decades of study on their sonar ability, it still wasn't clear how their brains processed this information to construct a mental map of their marine environment.

According to Thomas Nagel, the whole enterprise would be pointless, since the dolphin was so unlike us that we could never know what it was like to swim in the sea or echolocate. I thought otherwise. If we could figure out how the dolphin's brain worked during echolocation, we would be one step closer to understanding the dolphin's subjective experience. Besides, humans had rudimentary echolocation skills. Dolphin brains might actually tell us something about humans.

Lori said, "This is great! Nobody has done DTI on dolphin brains."

Sufficiently excited, I plunged in with the real question. "Do you know where we might get some dolphin brains?"

"Of course. I still have all the brains I scanned a decade ago. You're welcome to them."

This was better than I had expected. But the age of the specimens worried me. DTI on dead brains was hard enough. Who knew what kind of signal we could get out of a brain that had been sitting in formaldehyde for over ten years?

While I was pondering the complexities of the enterprise, Lori said in a lowered voice, "You know I'm leaving."

Lori had bounced around departments at Emory, but so had I. When you do research that doesn't fit neatly into a category, it's often difficult to find a home. It's not that fellow academics are hostile. Sometimes it's just a matter of having someone to talk to.

Lori's involvement with animal advocacy had consumed more and more of her time. The issue came to a head when she learned of the annual dolphin slaughter in Taiji, Japan. Lori became an advocate for whales and dolphins and later had a prominent role in the 2013 film *Blackfish*. When it reached a tipping point, she had to choose whether to go all-in as an advocate for dolphins and other animals or to continue teaching basic neuroscience to undergraduates and trying to eke out a research program. The choice was clear.

It was only then that I noticed the boxes in her office. She was leaving in a week.

"Where are you going?" I asked.

"Utah."

This didn't make sense.

"It's a place with special meaning for me. I'm setting up a new foundation called the Kimmela Center for Animal

Advocacy. We're going to bridge the gap between animal research and animal advocacy. You shouldn't have to choose one
or the other."

I admired Lori. It took guts to make such a radical change.
And although I hadn't entirely come around to her way of
thinking, that was just a matter of time.

According to Lori's instructions, the brains could be found
in the basement of one of the dreariest buildings on campus. Everyone called it the old dental school because it had
housed Emory's dental school until the school closed in 1990.
In recent years, the dental school's anti-Semitic practices had
received a public airing, and there had been an official apology
in a documentary confronting the university's past. And while
institutions can apologize for historical wrongs, they can never
quite erase the association from the location in which they occurred. If ghosts existed, the old dental school surely had a few
lurking in its corridors.

Peter and I went looking for the dolphin brains. The brain
room was unmarked. Had Lori not told us their precise location,
we would never have found them. We heaved open the door
and flicked on the lights.

What used to be a lab was now reduced to storage. Lab
benches were piled with boxes, and a thin layer of dust had settled over everything. A fume hood, meant to trap dangerous
gases or noxious odors, was cracked open, although I did not
detect the sound of air flowing that a working fume hood should
make. A dozen plastic buckets sat inertly atop each other inside the hood. The buckets had yellowed with age, and each was

filled about halfway with a dark brown fluid. Paper labels, taped to the sides years before, hung by mere molecules of adhesive.

"This must be them," I said.

Peter nodded.

We hadn't thought to bring gloves, but we had to see what was in the buckets. I lifted up the door to the hood and picked one of the buckets from the stack. Behind the discolored plastic, I could see something was submerged in the fluid.

I pried open the lid and was hit with the smell of formaldehyde. Even through watering eyes I could see a large brain. It didn't look like any other brain I had seen. It was big—at least half again as big as a human brain. And it was round like a soccer ball. The label read "Tursiops." We were staring at the brain of a bottlenose dolphin. Apart from the disgusting brown fluid, it looked to be in pretty good shape, especially considering that it had been sitting in the stuff for more than a decade.

"This isn't too bad," I said. "We need to change the formaldehyde, but at least it's not a pile of mush."

We opened each of the containers and noted which specimens looked the most promising. Doing MRI scans on postmortem specimens this old was uncharted territory. We would soon find out if we could measure anything useful about how dolphins' brains were wired.

Peter and I loaded a half-dozen of the buckets on a cart and wheeled them back to the lab, where we could get a good look at them. The brains in Lori's collection ranged in size from grapefruits to soccer balls. The bottlenose brains were the most impressive. Not only were they big, but they contained more intricate folding patterns than a human brain. More folds meant more surface area, which meant that even more neural tissue was buried in all those grooves, but none stood out as a central sulcus, the key landmark of the primate brain separating the frontal lobes from the rest of the brain.

My initial enthusiasm began to wane as I realized the complexity of the task before us. If we couldn't identify basic parts of the dolphin brain, how would we figure out how it worked?

In terms of frontal lobes, dolphins look more like carnivores than primates. When we took a closer look, we quickly saw that dolphins had a cruciate sulcus, but it was located so far forward that barely 10 percent of the brain could be called frontal. This was perplexing. For all the intelligence that dolphins possessed, how could they have such small frontal lobes? Either the cetacean anatomists were wrong, or we knew nothing about the parts of the brain responsible for cognition.

We had four bottlenose brains and two smaller ones—a common dolphin and a spotted dolphin. Bottlenose dolphins had a reputation for being the most intelligent and social of the dolphins, so these brains were our first choice to scan. Although they looked to be in pretty good shape, the important stuff was on the inside. A quick MRI scan would tell us if the interiors had liquefied.

Peter lifted each of the bottlenose brains from their buckets and let the excess liquid drip back in a viscous stream. We double-bagged each of the brains in zip-top bags and headed to the scanner, cradling the brains like macabre rugby balls.

I picked the best looking of the bunch and laid it in the MRI head coil. Even though it was bigger than a human brain, the absence of a skull allowed it to nicely fill the space. Peter stuffed some foam around it to keep it from moving, and we sent it into the center of the magnet.

I ran a quick localizer scan.

"This looks good," I said. "We can see gray and white matter throughout."

I was getting excited. These brains had been sitting in buckets for more than a decade and yet they were still giving off signals we could detect. Hopeful, I set up for a high-resolution structural sequence and pressed *Scan*.

The scanner initiated its pre-scan routine. A series of clicks and whirs could be heard deep inside the machine as the software compensated for the foreign shape now in its interior. After thirty seconds of adjustments, the actual scan was heralded by a loud buzzing. Normally, this type of scan would take about two minutes, but because the bottlenose brain was so big and round, it would take five to cover the whole thing. All we could do was wait.

After what seemed much longer than five minutes, the scans popped up on the screen. Starting at the bottom of the brain, I paged through the slices.

"There's the cerebellum," Peter said. "Amazing detail."

Indeed. It looked like a finely reticulated fern.

When we hit the cortex my heart sank. It was full of holes. You didn't need to be an expert in brain anatomy to see that there was something obviously wrong with this specimen. The prospect of tracing connections through this Swiss cheese looked pretty dim.

We scanned the other two bottlenose brains, and they all had the same degeneration in the interior. I guess it wasn't too surprising, given the age of the specimens. The holes were in a zone equidistant between the outer surface and the ventricles. When the brains were extracted and immersed in formaldehyde,

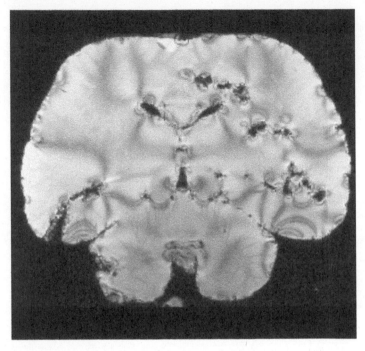

Brain of the bottlenose dolphin full of holes. (*Gregory Berns*)

their large size was an impediment to the preservative getting deep into the interior. It probably never preserved very well and just deteriorated over time.

That left the common dolphin's brain. About half the size of the bottlenose, it was round and compact, but quite a bit larger than a sea lion brain. We placed it in the head coil and scanned it just like the others. Much to our delight, it looked pristine. No holes! We set up the DTI sequence and left it to run overnight.

Because the dolphin brain was larger than the sea lion brain, the analysis of the data took almost a week. To make sure we had done everything correctly, we first examined the vector fields. The dolphin brain looked so different from the brains of terrestrial mammals that it might as well have been an alien from outer space. Everything had been jammed up into a big ball, and the usual landmarks were displaced into strange locations. Even the corpus callosum looked weird. For a brain of this size, it was uncommonly thin, meaning there were relatively few connections between the hemispheres.

To help get our bearings, we rendered the vector fields in three dimensions. This stripped away everything in the brain except the white matter, leaving a finely detailed map of all the connections. Because they are embedded in a spheroid object, the pathways are hard to see without software to do a virtual fly-through. So that's what we did.

As the animated image spun around, a sense of awe came over me. I got the scientist's tingles, knowing that I was looking at something that nobody had ever seen before. We could see the fibers connecting the cortex to the brainstem, running vertically and colored blue. As the brain spun around into a face-on view, the wings of the temporal lobes fanned out, and we could see all the cranial nerves branching off of the brainstem in tendrils of red.

Peter said, "I guess it worked."

All I could do was nod.

Satisfied that the diffusion data looked good, we decided to turn our attention to a scientific question. The dolphin brain was so different from anything we had seen before that Peter and I had a hard time deciding where to begin. Even the sea lion brains were easier. Because sea lions spent a lot of time on land, their brains were more like those of terrestrial mammals, and not that different from what we were used to seeing in the dog brains. The dolphin brains were an entirely different story.

A 3D rendering of the white matter tracts in the dolphin brain. This is a side view, with the nose pointed to the left. (*Gregory Berns*)

Echolocation is not as foreign as one might expect. Echolocation has two components, both of which humans possess: sound production and hearing. Dolphins are obviously more adept in its use than humans, but in the twenty years before Peter and I began our project, much had been learned about how dolphins performed echolocation, and it was far less mysterious than had once been thought.

Echolocation is the biological equivalent of sonar (a word that derived from "sound navigation and ranging"). The concept is simple to understand. A sound is emitted, reflects off underwater objects, and returns as an echo. The return time of the echoes reveals the distance to the object, and the sound of the echo relates to the size and texture of the object. Artificial sonar systems have achieved an astonishing level of sophistication, but they do not yet approach the accuracy of dolphins' echolocation abilities. Unsurprisingly, the US Navy has been intensively studying dolphins for decades.

Dolphins and whales generate sound like we do, with air, but with a major twist. We, like all terrestrial mammals, produce sound by vibrating the air in the larynx. During exhalation, a column of air passes through the vocal folds, which open and close, producing a series of air puffs. These bursts of air are further shaped by the throat, tongue, and lips. Dolphins have a larynx and vocal folds that can produce sound, but they are not the predominant mechanism dolphins use for sound production. Instead, all toothed whales, which include dolphins, have a pair of structures beneath the blowhole, which are unceremoniously called the "monkey lips" because of their shape (technically the monkey lips / dorsal bursae—MLDB). When a dolphin pushes air through her blowhole, the monkey lips open and close, producing vibrations similar to those created by terrestrial mammals' vocal cords. But instead of coming out of the mouth, the vibrations are transmitted to a fat-filled pocket in the front of the head, called the melon. The melon acts as an acoustic lens, focusing and intensifying the sound beam. Dolphins produce an impressive range of sounds, too. They use clicks, which are short bursts of high-frequency sounds, for echolocation, but they also whistle and buzz, which seems to be for communication with each other.

The dolphin's clicks exist in a realm far beyond human hearing. A teenager can detect frequencies of up to about 20 kilohertz (kHz), but dolphin clicks have a predominant frequency of well over 100 kHz. Even dogs and cats can only hear up to about 40 kHz. The high frequencies, though, are critical to sound detection underwater. In air, sound travels at 340 meters per second (m/s), or 768 miles per hour (mph), but in seawater, the speed of sound is a blazing 1,500 m/s (3,355 mph). Terrestrial mammals localize sound by the difference in arrival time between the ears, but underwater, this time is so brief that low-frequency sounds effectively arrive at the same time. This is why, for humans, underwater sounds seem to come from all around. The ultrahigh frequencies dolphins use mitigate the problem.

Dolphins have ears, but their external canals are mere pinholes. They hear through their jaws using bone conduction. This is not as strange as it may seem. Humans can hear the same way. If you place anything that vibrates—a phone, for example—against the curve in your jaw, you will be able to hear the sound. The shape of the dolphin's head, apart from being streamlined for swimming, also focuses incoming sound waves on the flare of the jaw. This arrangement gives them maximal sensitivity to sounds directly in front of them.

Much of the research on echolocation in dolphins has demonstrated their discriminatory abilities. In one study, they could tell the difference between the thickness of aluminum spheres even when it differed by as little as 0.3 millimeters. The use of high frequencies enables such great acuity, but only because the dolphin's brain is also lightning fast. One way to test the speed of the system is to present two clicks. As the clicks get closer in time, eventually they will be perceived as one click. The switchover is a measure of the time the nervous system takes to integrate the incoming information. In humans, this occurs at about 30 to 50 milliseconds. In dolphins, it happens at 264 *micro*seconds—making dolphins over one hundred times faster than humans at processing sound.

Although how dolphins are able to discriminate sounds so close in time is still not fully understood, their brainstems likely perform a substantial amount of processing before the signals get to the cortex or even the thalamus. For example, orchestras tune up to the A-note above middle C. This note is designated A440 because its frequency is 440 Hz. But even the best musicians would have a hard time telling if they were slightly flat at 439 Hz without a reference note. When played together, the difference in frequencies is 1 Hz, which is called a "beat" frequency because you can hear a slight beating—in this case, once every second. Based on their integration time, we can deduce that dolphins can detect beat frequencies of up to about 4 kHz.

All of these hearing elements, which at first seem so alien, turn out to correspond to elements that are present in our own brains. Despite what Nagel believed, it was not so hard to know what it was like to be a bat or dolphin.

I n fact, rather than making dolphins unfathomable, echolocation made dolphins a perfect test case for determining what we could learn about an animal's subjective state from its brain. Anatomists have known since the mid-twentieth century that the auditory pathways in dolphin brains are large. But there was still much unknown about how dolphins used sound reflections to create mental images of their world—if "image" was even the right word to use.

You would need to know where in the cortex the auditory information went. In terrestrial mammals, the auditory nerve carries all of the acoustic information to the brainstem. From there, the auditory stream splits into two paths, one staying on the side that it entered, the other crossing over to the opposite side of the brainstem. The streams then pass through a series of nuclei as they ascend toward the thalamus.

Right before the auditory information gets to the thalamus, it is collected in the spherical nucleus called the inferior colliculus. In mammals, the inferior colliculus is large enough that the left and right colliculi form a pair of bumps on the back of the brainstem. The superior colliculus, which receives visual information, sits right above them, and the relative sizes of the inferior and superior colliculi can be used as a crude measure of the relative importance of auditory and visual information to the animal. Dolphins have a really big inferior colliculus, and from there, it is a short hop to the thalamus.

The thalamus sits at the geographic center of the brain, between the brainstem and the cortex. In humans, it is the size of a small plum. The boundaries of the thalamus are well demarcated from the cortex, and its internal structure consists of dozens of separate nuclei. The different nuclei serve as waystations between the cortex and other parts of the nervous system. Some of the nuclei receive inputs from the spinal cord, passing sensory information from the body to the brain. Others form recursive loops with the cortex, receiving information via the basal ganglia and cerebellum and feeding it back to the cortex, presumably to facilitate movement coordination. In primates, a big chunk at the back of thalamus, called the pulvinar, is devoted to vision. Auditory information lands in the medial geniculate nucleus.

The auditory pathways have been well mapped in humans and rats, but very little work had been done in dolphins when Peter and I started out. The Marine Mammal Protection Act, a federal law, strictly limits what scientists can do with dolphins and other whales, and it has appropriate curbs on invasive studies of dolphins' brains. In the early 1970s, Russian researchers had attempted to map out the dolphin brain by inserting electrodes and seeing which parts of the brain responded to sounds. They also attempted to map out the connections between the cortex and the thalamus in dolphins by injecting radioactive substances that tracked down axons.

These limited studies of the dolphin auditory pathways painted a strange picture. In every terrestrial mammal, the auditory information turns laterally from the medial geniculate nucleus and heads to the temporal lobes. In humans, for example, there is an area at the top of the temporal lobe, called Heschl's gyrus, that receives this information. You can even find a map of different tonal frequencies there. But the Russian studies of the dolphin brain suggested that the auditory information landed in a part of the cortex near the crown of the head, almost in the back of the brain, where you would expect the visual stream to go. One theory holds that because of the importance of auditory information, evolution has caused the dolphin's auditory cortex to expand and migrate into regions occupied by visual information in terrestrial mammals.

Peter and I didn't know where the dolphin's auditory cortex was located, but we could easily identify the inferior colliculi. They were hard to miss. Sitting atop the brainstem, the inferior colliculi were distinct spherical structures over a centimeter in diameter. As we had done in the sea lion brains, we placed digital seeds. Instead of locating them in the hippocampus, we placed them within the inferior colliculi and instructed the software to reconstruct the tracts emanating from them.

The tracts went in two directions. Going down the brain-stem, the simulation perfectly tracked back to the auditory nerve. This was a great validation of the approach. In the other direction, a fat pipe of fibers connected the inferior colliculi to the thalamus, again as expected. But from there, the fibers took a turn and headed laterally to the temporal lobes. Just like in the brains of terrestrial mammals.

When Peter showed me the images, my initial response was, "That's great! It looks just like the auditory pathway of every other mammalian brain."

But when we sent the images to Lori, she called me right away. "This does not match what's in the textbooks," she said.

"What do you mean?"

"The auditory tracts should be going to the crown of the head," she replied. "Not the temporal lobe."

I was confident in our results and said, "Maybe the text-books are wrong."

This was not just an academic debate. The textbooks sug-gested that dolphins had a mainline from the auditory system to the visual part of the cortex, an arrangement that was dra-matically different from what is in human brains. But our DTI results suggested a closer link. We had found an auditory path-way going to the temporal lobe, which suggested that instead of "seeing" sound, dolphins heard it, like we do. But echolocation is more than hearing. Echolocation is an active process, with both productive and receptive components. Instead of seeing with sound, dolphins "paint" with sound.

We had scanned just one brain. Tantalizing as the results were, we needed to confirm that they weren't a fluke.

We needed another brain.

The only other brain that looked to be in decent shape was a grapefruit-sized specimen that had belonged to a pantropical spotted dolphin. As the name suggests, pantropical dolphins are found in warm waters on either side of the equator. In the eastern United States, they follow the Gulf Stream, sometimes as far north as Maine. Along with the common dolphin, the pantropicals are the most abundant of the cetaceans. The population even survived mass killings in the eastern Pacific. From the 1950s through the 1980s, millions of dolphins were killed in netting operations designed to catch large numbers of tuna. Prior to the invention of net fishing, the spotted dolphin population was estimated at between 3 million and 4 million. Currently it is 500,000.

Between the sea lions and the dolphins, Peter had become quite efficient at setting the brains in agar and running the overnight scans. Within a few days, we had the results for the pantropical. The signal wasn't as strong as in the common dolphin brain, which meant that the diffusion measurements had been buffeted by thermal motion to a greater degree. Even so, when we placed virtual seeds in the inferior colliculi, the same pathway to the thalamus and temporal lobe appeared. If we considered the left and right sides of the brains separately, then the temporal auditory tract had been observed in four out of four hemispheres. The finding appeared solid.

Our DTI findings revealed the roadmap for the receptive side of dolphin hearing. In both of the brains we scanned, the major pathway from the thalamus went to the temporal lobe—not the crown of the brain. In most respects, the tract was quite typical of mammalian brains, and the landing point gave the location of the auditory cortex.

But where did the information go from there?

To find out, we simply placed another virtual seed in this newly identified auditory cortex and tracked from there. Now, the paths headed rearward and upward.

The map that emerged suggested two auditory regions: one in the temporal lobe, like in terrestrial mammals, and one near the crown of the head next to the visual cortex. Bats are the only other mammals that have a similar arrangement. Because bats are terrestrial echolocators, how they echolocate is quite well understood. Bats have a primary auditory area in the temporal lobe, just like we found in the dolphins, but also have secondary and tertiary auditory regions just above and behind the temporal lobe. In some bat species, these auxiliary auditory regions contain neurons that fire preferentially to different echo delays, which translates to a distance map. Some bats also have another adjacent region that responds to changes in the sound of the echo. These bats can alter the sound of their emissions, tuning them to whatever they are targeting. This is called frequency modulation, just like the term we use for FM radio.

What's most remarkable about the similarities between bat and dolphin brains is that bats and dolphins are not closely related. You would have to go back at least 80 million years to find a common ancestor. The closest terrestrial relatives of the dolphin are the even-toed ungulates, like pigs, cows, goats, and sheep. They don't appear to echolocate (although another ungulate, the hippopotamus, uses underwater clicks for communication). Here was a classic case of convergent evolution. Echolocation evolved independently in bats and dolphins, but because of the similarity of purpose, converged to similar solutions—one in air, the other in water. An examination of their genomes supports this picture. Genes related to hearing and vision in bats and dolphins are more similar than would be expected for animals separated by 80 million years.

I found it comforting that we could place one piece in the jigsaw puzzle connecting dolphins to terrestrial mammals. Some of the satisfaction came from the simple pursuit of knowledge. The tree of life was truly magnificent, and finding connections between the branches gave me some perspective on where I, as a single member of the hominid family, fit. Looking at the brains of these animals with sophisticated tools like DTI told us that we were more similar than different. Even something as foreign as echolocation was really not that unusual when you broke it down to its components. Nagel argued against reductionism, but it was through reduction to white matter pathways that we identified commonalities between humans and dolphins.

Some of the satisfaction also came from the light our work shed on philosophy. Philosophers liked to refer to *qualia*—the subjective experience of a thing, like the color red. Consider an orchid with a multitude of pink and purple hues. There aren't enough words in any language to accurately convey the experience of it. There aren't enough color words, and forget about trying to describe the smell. If you believe qualia arguments, then each of us is forever locked into our own perceptive world without any way to verify that we experience things in the same way that others do. With words alone, how could I know that my *red* was not your *blue*?

Anytime someone played the qualia card, you could be sure they would claim that no matter how much we knew about the physics of something, we would never be able to explain the experience of it. It didn't matter that the color red could be completely and accurately described by electromagnetic fields with a wavelength of 700 nanometers. Physics didn't explain *redness*.

But I thought it did. Physics and biology could explain things in ways that words could not. The 2015 Internet brouhaha over *The Dress* would seem to support this view. Was it blue and black, or white and gold? Armchair philosophers took it as proof that qualia was a real thing, and we each had our own versions of the dress. But the scientific explanation of the dress was more mundane. Randall Munroe posted on his blog, XKCD, a simple illustration showing how the background color of the image affected our perception of the dress. Like most biological systems, color perception relies on differences, not absolutes. Something is relatively blue or gold compared to something else. The most likely explanation of the dress was that people focused on different elements of the background as the reference. Reduced to these elements, the mystery of qualia vanished. In fact, you can change your own perception of the dress color simply by focusing on different parts of the picture.

It was no different with the dolphin brains. Echolocation was not some alien skill that we could never relate to. Echolocation was just an amplification of perceptual skills that humans also possessed. But so far, we had been studying perceptual systems. What about something deeper? Some cognitive process that bound us all together? Something social.

Chapter 7

Buridan's Ass

In the early days of neuroimaging, from about 1995 to 2005, fMRI exploded in popularity among cognitive neuroscientists. Unlike its predecessor, positron emission tomography, fMRI didn't require injections of radioactive isotopes. It was safe and fast. Imaging centers popped up at all the major research universities, at first in the radiology departments of medical schools, and then in psychology departments. For the first time, non-MDs were using imaging tools previously reserved for radiologists. The use of fMRI democratized neuroimaging. On the whole, I think this was a good thing. The more people who can use a technology, the more likely innovation and discovery will occur.

As the field adopted this new tool, it collectively spewed out thousands of one-off studies. It was common to see papers and media coverage showing a colorful picture of a human brain with a hot spot of activity for some arbitrary mental phenomenon. Emotions figured prominently, as in, "This is the happy part of your brain." I was guilty of it, too, at least by association.

My early research on parts of the brain associated with decision-making made the cover of *Forbes* with a lurid headline, "In Search of the Buy Button." Scanner time was expensive, so these early studies tended to have small sample sizes, usually twelve to twenty participants. But the small sample sizes of these early studies made it likely that many of the results were probably just noise. Very few of the results were replicated, mainly because nobody wanted to waste time and money confirming someone else's experiment when there were so many new and exciting things to discover about the brain.

Because of the lure of pretty brain pictures with blobs of activation, the harshest critics of neuroimaging called it blobology, a neologism derived from the centuries-old practice of phrenology, which attempted to decipher personality traits from bumps on the skull. To the critics, fMRI was just phrenology repackaged for a twenty-first-century audience.

But to those of us who were doing it, the early days of neuroimaging were thrilling. A cowboy mentality pervaded the field, and we had the feeling of boldly going where no one had gone before. It was just so easy to dream up experiments, gather up a bunch of grad students for subjects, and collect a slew of data over a few nights or weekends. The only question was whether you had access to a scanner and the money to pay for time.

The party ended for cowboy neuroscience sometime around 2008. First, Craig Bennett, then a post-doc at Dartmouth, shoved a dead salmon in an MRI. Bennett found that without proper statistical corrections, it appeared that the salmon had neural activity. This led to an immediate reconsideration of how the field was reporting results. Second, Ed Vul, then a graduate student at the Massachusetts Institute of Technology, circulated a paper provocatively titled "Voodoo Correlations in

Social Neuroscience." Vul showed that many high-profile fMRI papers had inadvertently run their statistics twice, in essence, double-dipping in their data and making their results look better than they actually were.

But science has the property of self-correction. Sometimes it takes a while, but eventually people recognize the mistakes of early experiments, and the standard for good practices evolves. With neuroimaging, sample sizes gradually crept up, although they remain woefully small. And because of folks like Bennett and Vul, the most influential change was the recognition of how likely we all were to get false-positives while investigating the brain. A false-positive occurs when a researcher reports a region as being active during a cognitive task when in fact he or she has only observed a blip in a sea of random fluctuations. It is not always easy to tell the difference, but at least the bar had been raised.

In many ways, the Dog Project was a return to the Wild West ways. When Mark and I began the project in 2011, a chill had already fallen over neuroimaging. Journals demanded larger and larger sample sizes and more rigorous statistical corrections. So when we had trained all of two dogs to lay still in the MRI, my colleagues were rightfully skeptical of the results.

But we persisted. We recruited and trained more dogs. We replicated our initial experiments with larger sample sizes. We improved our statistical methods to stay current with the field. And yet, one thorny problem continued to hound us.

Many scientists had continued to question our interpretations of brain activity. The criticisms were not restricted to the Dog Project. As part of the retreat from neuroimaging, Russ Poldrack, a neuroimager at the University of Texas at Austin, wrote an influential paper about the difficulty of deducing mental processes from brain activity. He argued that because the brain is so interconnected, individual parts may have more than one function. What an individual region is doing at any one time depends not only on its own activity, but also on what other connected regions are doing. Because of the brain's interconnectedness, Poldrack said, one can't infer a mental process from activity in a single region by itself. He called the problem "reverse inference."

There are a couple of solutions to the reverse inference problem. The first is to consider the coordination of activity across many brain regions when interpreting mental processes. This approach gave rise to the field of connectomics. In practice, the study of connections requires lots of data and a cooperative participant who will hold still in the scanner for long periods. As well trained as the MRI-dogs were, this type of study was beyond their capacity for patience. The second solution is to design better experiments that home in on a specific cognitive process. Sometimes this approach requires multiple experiments. Mark and Peter and I decided to take that path.

The harshest criticisms of the Dog Project came from the camp of hard-core behaviorists. Admittedly, few scientists still adhered purely to the principles formulated by Skinner, but behaviorism seemed to be a fallback interpretation for any of our findings.

In a variation of the reverse inference theme, the complaint went something like this: Weren't the dogs just chaining together associations? After all, we had trained the dogs to go in the scanner by first rewarding them for staying in the head coil, then for staying in the head coil in the mock MRI, then for staying there with noise, and ultimately, for staying in the real thing. Even in the Go-NoGo experiment, they had been rewarded for not moving. Maybe, from the dog's perspective, everything was a sequence that ended in a food reward. If so, then all the brain activations we observed could be accounted for by nothing more complex than associative learning. There would be no need to invoke thinking, or emotions, or even a mind!

To be fair, I had thrown down the gauntlet in the *New York Times*, in which I wrote, "By looking directly at their brains and bypassing the constraints of behaviorism, MRI can tell us about dogs' internal states."

Fighting words. I believed them when I wrote them, and I stood by them in the face of the behaviorists' criticisms. At the time, we had scant evidence that we were measuring emotions—really we just had a few dogs responding to hand signals in the scanner. By design, the original experiment was a Pavlovian conditioning experiment, and so it was easy to see why the behaviorists took issue with my interpretation of animal sentience. But I had lived with dogs my entire life, and I couldn't accept that these fellow creatures were little more than mindless automatons. They had personalities, and they had likes and dislikes, and purposeful behavior that suggested a higher level of thinking than behaviorist models gave them credit for.

As the Dog Project continued to grow, with more dogs participating in ever more complex experiments, we could no longer dodge the question of the dogs' motivation. Hand-waving explanations didn't get around the vagaries of reverse inference from brain activity. The time had come to tackle the question directly.

Was it really all about hot dogs?

After four years, I had gotten to know the MRI-dogs as well as I knew my own dogs. They all had individual personalities. Some were playful. Some were shy. Some, like Libby, became agitated with strange dogs. Others didn't care. And while they all loved food, some, like Kady, seemed to care more about what their humans thought, at least when the humans were nearby. Many of the dogs would gladly turn down food to play a game of fetch or tug-of-war.

Obviously, we couldn't play with the dogs in the scanner, but the owners could certainly offer up effusive praise. This was going to be about the value of a social reward for its own sake. Would it be as rewarding as a morsel of hot dog? We called the experiment *Food vs. Praise*.

Movement presented a major challenge in this experiment. There would be no way to compare food and praise directly, because the dogs moved when they ate the treats. Somehow, the dogs would have to hold still for both types of rewards. We spun our wheels for several weeks as we tried to solve this puzzle. We considered holding a treat in front of the dog's nose, as we had thought about for the marshmallow test, but even if the dogs didn't move, that would still result in a sensory confound. Because there is no smell from verbal praise, we might just end up comparing the smell of food to its absence.

What we really wanted was an experiment that stripped away all the sensory differences between food and praise, leaving us with the naked reward value. For weeks, it seemed impossible, and then, like a bolt of lightning, the solution presented itself. It had been inherent to the Dog Project all along.

From the beginning, we had used hand signals to indicate a variety of outcomes. One raised hand meant: *Hot dog.* Two hands pointing toward each other meant: *No hot dog.* Arms crossed meant: *Don't move even when you hear the dog whistle.* The key to getting interpretable fMRI data was to focus on the response to the hand signal, not the thing that followed it, which was usually corrupted by motion artifacts. Brain activity during the hand signal measured the degree of anticipation. In humans, it was already known that a key brain structure, called the caudate nucleus, responded in anticipation to things that people liked, such as food, money, and music. So when we discovered in our early experiments that the dog's caudate reacted similarly to hand signals, in anticipation of treats, we knew we were on to something important.

Anticipation is extremely interesting because it is internal to the dog's mind. There is no direct behavioral manifestation of anticipation, so it is a perfect use of brain imaging to understand a mental process that would otherwise be inaccessible. We humans know what anticipation feels like, although admittedly it is hard to describe in words. Anticipation of something good is thrilling, sometimes even better than the thing itself, while anticipation of something bad is dreadful, often worse than the outcome. Think dentist.

Hand signals had served us well in many experiments. Because the owners were always in view, the dogs focused their attention on their human. Hand signals kept them engaged and motivated to stay in the scanner, because that's where their human was. But we couldn't use hand signals to study food vs. praise. If the owner was in view the whole time, the dog might get a continuous stream of social reward, and this would not be a fair comparison to food reward, which was intermittent. Plus, the owners normally delivered the treats, so the dogs were getting both food and social rewards during that interaction.

There was no avoiding it. The owners would have to be out of view of the dogs for this experiment. To be comparable to the food reward, the human would have to periodically pop into view and praise the dog. But we still needed to create a state of anticipation for both types of reward.

If the owner had to be out of view, the only alternative was to present new visual cues, one that indicated food, and another, praise. Projecting images onto a screen at the end of the magnet—say, on a computer monitor—was one possibility, but that would make it impossible for the owner to interact with her dog in person. Besides, none of the dogs had shown any interest in TV or surfing the Web at home (although that might have been due to lack of species-appropriate content).

One evening, something in the yard caught Callie's interest. I knew this because Cato started barking incessantly. That, in itself, was not unusual. Cato always barked. It was his catchall form of communication. The only thing that varied was how much energy he put into it. When one of the other dogs had something of interest, Cato's barking became louder and more frenetic.

I knew Callie had something good because Cato was barking at fever pitch. That usually meant she had caught a small animal, probably a mole.

But before I could go outside to see what she had, a scrum of dogs burst through the kitchen door. Callie was in the lead, proudly carrying not a mole but a dirty Barbie doll—no doubt a relic of my daughters' childhood. Barbie had seen better times.

Now, as Callie's trophy, she presented the solution to the food vs. praise experiment.

We would use toys.

The next day, Peter and I brainstormed about the details.

"How do you think we should present the toys?" Peter asked.

"On sticks?" I replied. "We could glue a colorful toy on the end of a wooden dowel."

"Right. Then we just hold up each toy for ten seconds and follow it with the reward."

I liked this idea. "One toy would signify impending hot dog, and another would mean the owner will give praise."

Peter scrunched up his forehead in thought. "The owners would still need to be out of view while the toys were being displayed, at least until they praised the dog."

He was right. For this experiment to work, we would have to train an association between each toy and the appropriate reward. Would the dogs stay in the scanner while watching a puppet show of toys appearing in front of them? There was only one way to find out.

Peter and I went home with the mission to scour our kids' toy bins for colorful objects. At first, I thought it would be fun to use Callie's newfound Barbie, but I quickly realized that she would not look good mounted on a stick. Instead, I found a sharp-looking medieval knight atop a horse. The toy was about six inches long by four inches high and brilliant cobalt blue—a perfect color for hitting the peak sensitivity of the blue receptors in the dogs' retinas. Even if they couldn't make out the man on a horse, they would easily be able to distinguish this toy from one of another color. While rummaging about, I found one of the kids' hairbrushes. It was one of those cylindrical designs with spiky things protruding all around. Both objects appeared to be free of metal, which was necessary to keep them from being sucked into the magnet.

Peter snuck away a toy car from his son. It was a toddler-friendly version of a VW bug in eye-popping magenta with yellow wheels. It would hit the dog's other chromatic receptor, which had peak sensitivity to yellows.

With a little work on the drill press, I bored holes in all the toys and epoxied them to three-foot wooden dowels.

Kady, the placid Lab mix, was the first to try the new experiment. At the practice facility, we set up the mock MRI and positioned the steps at the base. Kady's owner, Patricia King, placed the head coil in the tube and said, "Kady, coil!"

Kady trotted up the steps and settled into her chin rest.

I set a camera on a tripod, aiming down the bore. Except for the brief moments when Patricia popped back into view to praise Kady, there would be no humans in view. We needed the camera to capture Kady's reactions.

This was going to be a two-person operation. I sat on the floor beneath the MRI tube, out of Kady's sight. A computer had pregenerated a random ordering of the three trial types. The blue knight would signify impending hot dogs. The pink car meant praise. And the hairbrush was there for a visual control and meant nothing.

For the first pass, I cycled through the stimuli quickly, holding up each toy for about three seconds. After the blue knight, I skewered a piece of hot dog on the end of a long dowel, which one of my daughters had dubbed the treat-kabob, and maneuvered it like a pool cue up to Kady's mouth. I could hear her greedily sucking it down. After the pink car, I would nod to Patricia, and she would pop into view, exclaiming, "Kady! What a good girl!"

After ten repetitions of each trial type, we took a break and played back the video. Kady was inscrutable. She stayed planted in position throughout the whole thing, and I couldn't read anything into her expression that would tip off what she was thinking. In some ways, that was a good thing. Otherwise, there would be no reason for doing the brain scanning. Just to make sure Kady learned the association of the toys with the rewards, we went through the repetitions two more times. Then, to make sure everything stuck, we repeated the whole session two weeks later, and Kady was ready for the real thing.

Kady did just as well during the MRI scan as she had in the training sessions. She excelled at these passive tasks, in which she didn't have to do anything except stay in the head coil. This was why Kady was our go-to dog for first runs. Confident that this task was well within the capabilities of the other MRI-dogs, we taught all of them the meanings of the three objects just like we had with Kady. Within a few months, we had completed their scans.

The analysis of the fMRI data would be straightforward. As in our original experiments, we would focus our attention on the caudate nucleus. The caudate forms a large inverted "C" arching from the front of the brain and trailing out almost to the rear. In humans, the front of this arch is connected to the parts of the frontal cortex associated with cognitive functions like planning, or, as we saw with the Go-NoGo experiment, impulse control. Farther down the arch, the caudate merges into a structure called the nucleus accumbens, which is connected to systems associated with reward and motivation. We would measure the activity in this part of the caudate during the presentation of the different objects. The key comparison would be between the blue knight and the pink car. If one object elicited more activity than the other, then we could conclude that the anticipation for the corresponding reward was more intense.

As a check that the dogs understood that two of the objects had meaning, we first compared the activity of the knight and the car to the hairbrush. Reassuringly, both the knight and the car evoked strong caudate activity while the brush did not.

But when we compared the two to each other, there didn't seem to be a significant difference. The caudate activity increased for both the knight and the car. This increase indicated that the dogs had learned that the knight meant food and the car meant praise, but they liked and anticipated both equally.

This was a modestly interesting finding, but not the home run we had hoped for. In the end, we scanned fifteen dogs—a record number for the project and plenty to satisfy the referees who would scrutinize our findings.

Each day, Peter and I would stare at the map of brain activity. There wasn't even a hint of difference between the food and praise cues.

"Maybe," I said, "some dogs like food and others like praise."

Peter finished my thought. "And by lumping all the dogs together, any differences average out."

In the cognitive control experiment, we had found a strong correlation between frontal activation and a dog's ability to exert impulse control in a variety of contexts. So even though we had gained statistical power in the food vs. praise experiment by pooling the activations in the larger sample size, we had not yet probed the differences between the dogs. As in the marshmallow test, one size did not fit all.

Peter extracted the caudate activity for each dog and ranked the dogs by the magnitude of the difference between food and praise. Pearl, Kady, and Velcro had the highest differential in

Pearl (*front*) and Ohana (*rear*) after food vs. praise scanning. (*Helen Berns*)

favor of praise, while Big Jack, Truffles, and Ozzie swung the furthest toward food. Big Jack had earned his nickname for a reason. Let's just say he wasn't born "Big." Ozzie was a cute Yorkshire terrier, and if there was one dog in the project who loved food, it was him. Although his owner, Patti Rudi, doted on him, she was under no illusions about his motivation. "It's all about the food for Ozzie," she once told me.

At the other end of the spectrum were the dogs who stuck closest to their owners. Pearl, Kady, and Velcro were all sweet dogs, and while watching them train, I couldn't help but believe they cared most about what their humans thought. Food was almost incidental.

The ranking of caudate activation seemed to roughly line up with the dogs' personalities, but we needed a less subjective way to measure this.

Usually we think of preferences as a human phenomenon, but animals have them, too. The measurement of the dogs' preferences could be approached in several ways. If we had been dealing with humans, we might simply ask a person to make a choice. The Pepsi Challenge, which pitted Coke vs. Pepsi in a series of commercials beginning in the 1970s, was typical of a forced-choice test. We needed a Pepsi Challenge for the dogs. Instead of Coke vs. Pepsi, it would be food vs. praise. But because dogs can't tell us their preferences, we would need a test in which their behaviors revealed their preference.

Choice tests are used in a wide range of animal experiments. Primates, like monkeys and apes, can be readily taught to point to or tap objects, while rats and pigeons can learn to press bars for different outcomes. Although we could have taught the dogs in the project to nose-poke a target, the logistics of delivering the two rewards presented some challenges. Delivering food could have been done easily through a chute of some sort, but human praise required a different type of delivery system. We thought about hiding the humans behind a door and having them pop out to praise their dogs, but the training facility wasn't set up for anything like that.

In the interests of simplicity, we settled on a maze.

In the main training room, we linked together a bunch of baby gates, creating a barrier in the shape of a V. The vertex was positioned in front of a door, behind which the dog was waiting. When the door was opened, the dog would enter the main room. Faced with the V, the dog would have to make a choice as to which way to go. At the end of one arm, the owner would be seated in a chair. To avoid encouragement with facial expressions, she would have her back to the dog. At the end of the other arm would be a food bowl with a treat inside.

This was an unusual situation for the dogs. We would have to familiarize them with the setup if the maze was going to help us measure their preferences.

Pearl, the energetic golden retriever, was the first dog to go. On the first four trials, Mark led Pearl from the door to the end of each arm for two trials of food and two of praise from the owner, one of each type on the left and right. Then came the free-choice phase. Unlike the Pepsi Challenge, in which subjects picked only once, we had already decided that the dogs would need several opportunities to express their preference. In part, this was because we didn't know if the dogs would understand the task. It might take several trials before they understood what was going on. We figured that twenty trials should be enough.

On the first free-choice trial, Mark opened the door, and Pearl looked down the right arm at her owner, Vicki D'Amico. Pearl then looked left, did a double-take toward Vicki, but finally trotted happily to the food bowl. The whole process took five seconds.

Vicki switched locations with the food bowl. We had anticipated that some of the dogs might develop a side preference, so it was important to alternate the location of food and owner. On the second trial, Pearl headed straight to Vicki. On the third trial, she finally went to the right, where Vicki was located. After twenty trials, it was apparent that Pearl had a bias to the left (75 percent of the time), but she also went to her owner 70 percent of the time. On a handful of trials, Pearl couldn't decide what to do. She just wandered around the room.

Within a few months, we had tested all the dogs on the preference test, and they were all over the map. Some dogs chose food 100 percent of the time, while others chose their owner on as much as 85 percent of the trials. Most dogs were somewhere in the middle.

But there was more to this than the simple percentage of choices. In the early trials, many of the dogs made a beeline for their owner, seemingly forgetting that food was available just over the barriers. Sometimes they would get halfway to their owner and catch a glimpse of the bowl, but, unable to figure out how to get to it, continued on to their owner. Certainly the people were pleased when their dogs chose them over food.

Eventually, most of the dogs figured out that food was there for the taking and it was okay to go to the food bowl. Sometimes you could see the light bulb click on, as if the dog were thinking, "Hey, this is great. I can have food and still see my human!" (In hindsight, we should have made the barriers opaque.) Some dogs, like Big Jack, caught on early. Jack went to his owner, Cindy Keen, on the first three trials, but on the fourth he realized he was missing out on the treats and went to the bowl. After that, he sampled back and forth, settling on a ratio of about three trips to the food for every one trip to Cindy. In contrast, Velcro remained true to his name. It wasn't until the sixteenth trial that he strayed from his owner and went to the food. And even then, you could see he struggled with the decision. Kady was similar. Consistent with her cautious personality, it took her seventeen trials to go to the food.

Simply counting the number of food and owner selections failed to capture the complexity of a dog's decision-making process. The problem was that our conception of preference was wrong. When we had designed the experiment, we had assumed that the proportion of selections would reveal how much each dog preferred food or praise. Since the 1960s, scientists had characterized animals' preferences by the ratio of their responses to the choices offered. It was called the matching law, and was originally demonstrated in pigeons. The matching law stated that an animal would allocate its responses in proportion to how much it liked the outcome. Pigeons would peck more at a lever that released more food than at a lever that released less. The ratio of pecking at the two would match the ratio of food released. Rats behaved similarly. At first glance, so did the dogs.

Not to diminish the cognitive skills of pigeons and rats, but dogs are more complex. They care about things other than food—like the approval of their humans. Inadvertently, we had placed them in a situation that could have turned them into Buridan's asses.

Jean Buridan was a fourteenth-century French philosopher who used the ass to illustrate a paradox of free will. In the paradox, a donkey that is both hungry and thirsty is faced with a pile of hay and a bucket of water. Unable to decide which to choose, he dies of hunger and thirst. The failure to choose would be taken as evidence of a lack of free will. Some have argued that the ass is entirely fictional and that there is never a situation where two options are so exactly equal that an individual is frozen in indecision.

I didn't think Buridan's ass was just fiction. On any given trial, the dogs, to varying degrees, wanted both the food and their human, but they could have only one. We humans know exactly what that feels like. Choices concerning our spouses, the colleges we attend, and our careers can paralyze us with indecision. I have seen many friends and colleagues procrastinate over major life decisions to the point that the options eventually disappear. It isn't so much a problem of equally appealing choices, but rather, one of fearing the regret associated with a decision. But, in my view, pathologic procrastination is the very definition of Buridan's ass.

It seemed like the dogs were experiencing something similar. After all, some of them had been unable to make a choice on some trials, and had chosen to wander around the room instead. Although our original plan was to focus on the proportion of food and owner choices, it became clear that the sequences themselves contained important information about the way the dogs resolved their internal paradoxes.

Jack's sequence went something like this: food-owner-food-owner. In contrast, Ohana and Pearl, while they had a similar proportion of responses, followed a different sequence: food-food-owner-owner. Jack was a sampler, while Pearl and Ohana tended to stick with something for longer stretches.

The simplest way to analyze a sequence of choices is to break it down into sequential pairs. These are called state transitions, and in our experiment we had four types: food-food, food-owner, owner-food, and owner-owner. By counting the number of these transitions, we could calculate the probability that a dog would choose food or owner based on the prior choice. Stickiness showed up as a high proportion of either food-food or owner-owner sequences. The difference between the two told us how much the dog preferred food or praise and the strategy the dog used to resolve the paradox of choice.

The stickiness measure seemed to match up with the dogs' personalities. Velcro had a high stickiness for praise—82 percent—but zero for food. Jack was another story. While Cindy had unshakable faith in his love, Jack also loved his food. His stickiness for food was 62 percent, but it was only 43 percent for Cindy. Jack was a player.

When Peter and I compared the V-test to the brain data, a relationship popped out. The stronger the caudate response to the owner, the stickier the dog was to the owner on the choice test. The fMRI data seemed to tap into the relative love each dog had for food and human.

While this may not seem surprising, demonstrating links between brain activity and behavior, especially outside of the scanner, was devilishly hard. The contexts were completely different. In the scanner, the dog simply lay still, watching inanimate cues and waiting for either food or praise. The caudate response captured something about the relative state of anticipation. But in the free-choice test, many factors came into play, the most important of which was the simultaneous availability of both options.

In standard economic theory, the availability of several potential outcomes shouldn't affect the amount of enjoyment that an individual gets from a chosen reward. Expected utility theory assumes that individuals make choices to maximize the future benefit. After all, if I like vanilla ice cream, what difference does it make if I have to choose between vanilla, chocolate, or pistachio, all of which I like? Presumably, whatever I choose is what I most desire at the moment. But even though expected utility theory is foundational to all modern economic analyses, it often fails to capture the internal psychology of decision-making.

Of course, not all decisions turn out the way we expect. Disappointment is a key driver of learning in the brain. And disappointment leads to regret—the knowledge of what might have been. With ice cream, no matter what I choose, I still wonder whether I would have liked the other flavors better. By the 1980s, however, some economists had formulated alternative theories of decision-making that incorporated the possibility of future regret. Regret theory said that people sometimes choose less desirable outcomes just to avoid the possibility of regretting something in the future.

On the surface, regret would seem to be a uniquely human experience. Its very nature depended on the recognition of alternative realities, which is a complicated feat of imagination. For regret to factor into decision-making, you not only have to experience regret, but anticipate that you might experience it. In 2004, the human orbitofrontal cortex was identified as a critical area for regret, because people who had suffered strokes to that region no longer experienced it. So it was somewhat of a shocker in 2014 when David Redish, a neuroscientist at the University of Minnesota, reported on the neural basis of regret in rats.

Redish developed a Restaurant Row for rats. This was a circular maze with four spokes, and at the end of each spoke a different flavor of food was dispensed: banana, cherry, chocolate, or unflavored. When a rat entered a quadrant, an auditory countdown started, which was indicated by a tone decreasing in pitch every second. If the rat stayed until the end of the countdown, he received the flavor in that zone. If the rat moved on, the countdown for that zone stopped, never to begin again. At that point the rat would have to get some other flavor. Redish thought that if a rat liked some flavors more than others, he would spend more time in those zones waiting for the clock to tick down. But because the length of each countdown was random, it was always a choice between waiting for the preferred flavor or moving on to something less desirable. Restaurant Row was very much like Mischel's marshmallow test.

The most regret-inducing outcome occurred when a rat,

being particularly impatient, chose not to wait for his favorite flavor. The rat moved on only to discover that the next choice necessitated waiting even longer for a less desirable outcome. In these circumstances, the rats often looked back to the forgone option. And Redish found that during those look-backs, the neurons in the rat's orbitofrontal cortex and caudate nucleus were highly active. The inescapable conclusion was that rats, like humans, simulate what might have been.

If rats experience regret, it seems very likely that dogs do too. Our MRI data for dogs pointed to the pivotal role of the caudate nucleus in signaling the value of the anticipated outcomes. Although we didn't give the dogs a choice in the scanner, we can assume that the caudate, which is also connected to the orbitofrontal cortex, would carry information about might-have-beens. Just like in human brains.

Admittedly, it is hard to conceive what it is like for a dog or a rat to experience regret. The data, though, clearly showed that they were capable of running simulations in their brains of what might have been, which is consistent with the fourth principle of brains: *Brains simulate possible actions and future outcomes so as to make the best possible decision for the situation at hand.* Just because humans have a word for it and animals do not doesn't mean that animals without language can't experience regret. On the contrary, the neural evidence suggests that they do. But this finding raises a question about the extent to which language is necessary for the subjective experience itself.

Besides, it is not entirely correct to say that animals don't have language. Animals may not talk, but those that live around people clearly understand some aspects of human speech. The important question is: What do words mean to an animal?

Chapter 8

Talk to the Animals

"Callie!" I said, "Get hedgehog!"

She waggled her whole rear in excitement and looked at me with an expression that I took to mean: *This is fun!* She spun around and trotted over to the lineup of objects I had placed on the floor across the room. Callie passed by several coffee mugs and nudged the spiky stuffed animal that I had dubbed "hedgehog."

I pushed the handheld clicker, indicating that she had picked the correct object, and Callie zoomed back to me to gobble up a piece of hot dog as a reward.

For the next trial, I wanted her to go to a rectangular piece of blue foam. I said, "Callie, get blue!"

She looked at me and then, wagging her tail, looked at the array of objects across the room.

"Callie! Go get blue."

She trotted over to the hedgehog and nosed it a few times. Not hearing the click, she looked back at me.

I did my best to remain expressionless.

After a few seconds, Callie gave up on the hedgehog. But instead of going to the foam block, she walked over to a door-stop poking out of the baseboard.

"Sorry, girl. That's not it." I motioned her to come back.

We tried it three more times. Despite having trained the word association to the foam block many times, she still wanted to go to the hedgehog. Maybe the hedgehog was just more fun to play with, but the way Callie was making errors seemed to say something important about how she processed language.

Although Callie was semiretired from the Dog Project, I still worked with her to debug new experiments. I was trying to teach her some simple word-object associations. And it wasn't going very well.

This particular experiment was inspired by the growing number of reports of dogs with large vocabularies. In 2004, a border collie named Rico was reported to have learned over two hundred words for different objects. And then, in 2011, John Pilley, a psychology professor, trained his border collie, Chaser, on more than one thousand words. Chaser created quite a buzz in the scientific community. If she could learn that many words, did that mean that all dogs had this capability? More fundamen-tally, did Chaser just associate arbitrary sounds with specific ob-jects, or did she have some rudimentary knowledge that a sound like "Binky" could be the *name* for a thing?

The question was one of semantic knowledge.

Did dogs have semantic systems in their brains? It would be remarkable if Chaser understood that words had meaning, and even more so if she understood that meaning could change depending on the context. Until recently, nobody had really thought to look. Most of the efforts in animal language had fo-cused on primates—chimpanzees and bonobos. But apart from a few individuals, there was not much evidence that even our closest relatives had much in the way of language ability or semantic knowledge. Part of the problem was knowing what an animal knew.

I wanted to do more than just learn what Callie knew. I wanted to talk to her in a way that she could understand. To talk to animals, three things would have to happen. First, we would have to figure out what they knew from human language. Second, we would have to learn to interpret their communication. And finally, we would need to create a system that allowed for two-way communication—something that translated our ideas into a form an animal could understand and, in the other direction, interpret an animal's way of communicating back to us.

It is not as far-fetched as it sounds.

In his most famous work, the French painter René Magritte painted a picture of a tobacco pipe. Beneath it, he wrote, "Ceci n'est pas une pipe." This is not a pipe. Magritte made explicit that it was a painting of a pipe, not the pipe itself. And so it is with our brains, which construct mental representations of the world with only the information we get from our senses as inputs. But our mental representations should not be confused with the things that are being represented.

It sounds confusing, but if we are ever to talk to animals, then we must solve the representational problem. When Callie looked at the stuffed hedgehog, her mental representation was different from mine—even though we were looking at the same thing. She saw the toy through the eyes and brain of a dog, and I saw it with human eyes. When I spoke the word *hedgehog*, I knew the word referenced the object. But Callie's behavior, especially with the blue object, suggested that she did not know that these words referred to things. I hoped that we could use fMRI to understand how words and objects were represented in her brain. That would be a step toward the first aspect of communication—knowing what she knew.

We based the idea on the groundbreaking work of Jack Gallant, a neuroscientist at the University of California at Berkeley. Gallant was using fMRI to decode the human brain. Initially, he had been interested in how the brain encoded visual scenes. Scientists had long known that the initial stages of visual processing extracted low-level features like contrast, edges, color, and motion. Beyond these properties, it remained a mystery how the human brain assembled visual images into a representation of a thing.

Gallant took a brute-force approach. He scanned human subjects in the MRI for hours while they watched movies. Later, he and his students hand-coded each movie frame for what was being shown. They then input the activity at each location in the brain at every second of the movie to a computer algorithm. With such a huge amount of data, the algorithm was able to find patterns of brain activity that reliably appeared when a participant saw certain images. The team was even able to reconstruct what the participant was viewing from moment to moment, based solely on his or her brain activity.

More recently, Gallant's lab took a similar approach with language. Instead of watching movies, participants listened to hours of podcasts in the scanner. Painstakingly, the researchers transcribed each of the 10,000 words in the podcasts into semantic representations, which is essentially a mathematical way of specifying what words mean. Again using a computer algorithm, they then determined how the meanings of the words were distributed throughout the brain. Apart from the fact that they could do this, the big finding was that they could identify clusters of meaning in the brain. Words that were associated with actions were located in one place, while words associated with quantity somewhere else, and words associated with social concepts in yet another place. The same word would often appear in different clusters, depending on its meaning, which varied by context.

There was no way that dogs had such rich semantic representations, but I suspected they had to have something. So the plan was to use Gallant's approach with a very small number of words. Actually, just two.

At first, all of the dogs did well. None of them had any problem picking out the named toy. But when we introduced a second toy, almost all of the dogs became confused. Like Callie, many of them would continue to go to the first object. Failing to get the praise or treat they expected, they would then fish for the right answer. We called it "bowling for treats." How was it that Chaser did this so easily, but the MRI-dogs had such difficulty?

Maybe Chaser was special. She was a border collie, and it was well known that border collies excelled at these types of tasks. In fact, the only MRI-dog in our program who seemed to understand this game was Caylin—also a border collie. But we weren't asking the dogs to learn a thousand words, just two, so all of us were surprised at how difficult it was to train the dogs.

Somewhere there was a disconnect. The dogs did well enough when the named object was in a field of distractors, but when both named objects were present, they did not do well. One possibility was that the dogs did not understand the word-object association, and they were just picking objects based on familiarity. Faced with two familiar objects, they just guessed, or went to the one they had learned first. Or maybe the objects weren't different enough. Maybe one fuzzy toy looked the same as another. To mitigate this possibility, we switched out the second toys for ones that differed in texture or squeaked. It was hard to say if this helped. Some dogs improved in their performance and others didn't. It was slow going for everyone.

Another possibility was that the dogs actually did understand the difference between the words but there was a problem with the way in which we were testing their knowledge. There weren't any consequences for making an incorrect choice, because they knew another trial would follow. So maybe the dogs just didn't have enough incentive to care about getting it right.

Another possibility was that we expected dogs to think like humans. I was intrigued by how Callie was making errors. Aside from her tendency to prefer the hedgehog, when looking for the block of foam she seemed to focus on the corner of the block. On several occasions it seemed like she was scanning the room for pointy objects. When she went into fishing mode, she would sometimes go to the doorstop. Other times she would go to the corner of a coffee table. I wondered whether, in Callie's mind, "blue" meant "pointy thing."

We humans take it for granted that a name refers to the whole object. But there is no reason to expect other animals to think like us when it comes to language. Dogs could be feature-bound where we humans take a gestalt view. The evidence was scant, but a few studies did support my idea that dogs mapped words to objects in a fundamentally different way from humans.

In 2012, Daniel Mills, a psychologist at the University of Lincoln in England who had published extensively on canine cognition, described how a single dog generalized from learned words. Again, the dog he used was a border collie. The dog was taught to associate a nonsense word (*dax*) with a furry object in the shape of a blocky U. Then, the researchers presented the dog with slightly different objects to see which ones he would choose as most similar. These objects varied in size, shape, and texture, but otherwise had similar characteristics. When humans do this task, they typically generalize to shape, a behavior that appears around age two. But Mills found that the dog he studied tended to generalize initially on size, and then later on texture, but never by shape. Size and shape are global properties of objects because they are defined by the whole thing. But texture is a local property, only discernible up close.

Beyond the question of global versus local properties, when I began my work with Callie and the hedgehog it was not clear whether dogs understood that words referred to objects. In most language tests, the words are nouns, which humans have no problem understanding as referring to things. Even two-year-old children get this. But it could be that when Callie heard "hedgehog," she interpreted it not as a noun but as a verb-object action meaning "get hedgehog." It may seem like a subtle difference, but if we are to communicate with animals, we need to know whether they interpret words as actions or things.

It is easy to teach dogs tricks. But tricks are actions. Teaching dogs that words could refer to things turned out to be much harder than teaching them to perform certain actions when they heard certain words. It may be that most dogs cannot understand that words can refer to objects. After all, the only way a dog can demonstrate knowledge of a word is to interact with an object in some way. In a dog's mind, a word may be a command to do something.

This should not be surprising. After all, animals have brains to do things. Even dogs, the first animals to live with humans, and the ones most likely to understand our language, evolved for action. They don't talk, and they don't read books.

Chaser is the only dog documented to have shown evidence of understanding the difference between a noun and a verb. Even with her enormous vocabulary, however, it seems unlikely that she understood anything more than simple verb-object pairings, because that is all the understanding she demonstrated. She had, in effect, acquired a rudimentary pidgin language.

Pidgin languages are common. They arise when people need to communicate with each other but don't speak the same language. Pidgins are grammatically simple, sometimes agrammatical, and contain a subset of words from the parent languages as well as mashups between them. Unlike in a full-fledged language, in a pidgin word order is often unimportant. Ray Jackendoff, a linguist, has argued that pidgin formed the basis for the evolution of human language. It's possible that dogs evolved a similar capacity during domestication.

Pidgin-level understanding is not limited to dogs like Chaser. Ron Schusterman taught Rocky, a sea lion, hundreds of combinations of action-noun pairs, and even modifiers for size and color. A bonobo named Kanzi also learned an impressive vocabulary of nouns and verbs, but he showed no evidence of understanding syntax.

Only dolphins so far have shown a capacity for understanding anything beyond pidgin-language. Beginning with John Lilly's groundbreaking attempts to communicate with dolphins, evidence has accrued for sophisticated language abilities in cetaceans. In the 1980s, Phoenix and Akeakamai, two bottle-nose dolphins, were taught more than verb-object relations. Their trainers showed that the dolphins understood word order and object attributes, suggesting that the dolphins treated words as symbols. Later work showed that Akeakamai also understood simple rules of grammar, or at least when rules of syntax were violated.

But how much meaning could we push into an animal's brain? Full-fledged language requires brain structures capable of learning not just a large vocabulary, but rules for stringing words together. Perhaps most importantly, language exists to exchange thoughts. For this to occur, individuals need to understand that words can represent actions, things, or both. The experiences with Chaser, Rocky, and Akeakamai showed that some animals could learn a lot of words, and even some rules of logic, but the process was laborious. To talk to the animals, we humans had to understand the limitations on the receiving end.

Take a name.

Names are nouns, albeit proper nouns, meaning they refer to specific individuals, organizations, and places. We use names to make clear who, out of all possible people, we are referring to. We also know that our own names refer to us as individuals, providing a label for our sense of self. Of course, we don't normally refer to ourselves in the third person, but when someone else uses our name, we instantly translate that into "I" or "me."

But how do animals treat names? If an animal doesn't have the faculty to understand that words are symbols, it is unlikely that they can translate their names into a sense of self. More likely, animals learn that a particular utterance means something interesting is about to happen and that they'd better pay attention. Whenever someone said "Callie," Callie directed her attention to whoever made that noise. I never got the sense that she equated her name with "me."

The experiences of animal trainers would support the attention-grabbing function of names. "Callie, sit," is thought to be more effective than "Sit, Callie." Humans don't have any trouble parsing the equivalence between the statements (although the first sounds more imperative than the second, which is closer to a request). Callie responds better to the first because her name gets her attention for the subsequent action. The reverse order requires her to remember the action that precedes her name.

Much of this may seem like common sense. But if we are to build a foundation for animal communication, we have to speak in ways that they can understand. I have written a lot about the similarities between the brains of humans and other animals, but when it comes to language, we must acknowledge fundamental differences.

I have come to the conclusion that communication with animals is possible, but only on a low bandwidth channel. When it comes to verbal communication, simple verb-object statements are probably the best one can hope for. Most nonhuman animals probably lack the brain resources to understand the difference between subject and object in a sentence. Although many animals have some sense of self, it is probably rooted in the physical domain. A dog like Callie has a sense of her body and an awareness of where her body ends. She wouldn't confuse her body with that of another dog. But it is unlikely that she can link her name to her physical or mental self.

Human semantic space is enormous. As part of their fMRI work, Gallant's team analyzed large bodies of text and determined which words and concepts tended to go with each other, and then they made maps of what semantic space might look like graphically. Gallant's map was a simplification. Even so, no animal could have a semantic space as rich as the one their map depicted. To make an equivalent map for animals, we would need to know what animals cared about and how these things were represented in their brains. Only where the space of an animal's concepts overlapped with ours would communication be possible. There would be no point in trying to tell a dog about a rough day at work when the concept of "work" was hopelessly foreign to him. He wouldn't have a semantic representation for "work" like we do. But could a horse pulling a hansom cab around Central Park have a representation for "work"? To answer this type of question, we would need a map, but it wouldn't be a linguistic map. A dog might have a semantic map like the one shown on the following page. This map is highly simplified, but it illustrates the space of human concepts and how it might overlap with a dog's space of concepts, while indicating that a dog might have concepts that we do not, or at least that we do not have labels for.

I should point out that I am being liberal with my definition of *semantics*. Many people would argue that semantics applies only to language. I take a broader view. The study of semantics is about the representation of knowledge. We use words to represent facts: *Humans landed on the moon in 1969.* But knowledge can come in different forms. The same knowledge can be represented visually, as in the old MTV logo of a man standing on the moon next to a flag. The point is that animals can represent and communicate knowledge in nonverbal ways.

One kind of knowledge that dogs can represent concerns emotional states, not only in other dogs, but in humans. In 2015, Ludwig Huber, who runs the Clever Dog Lab at the University

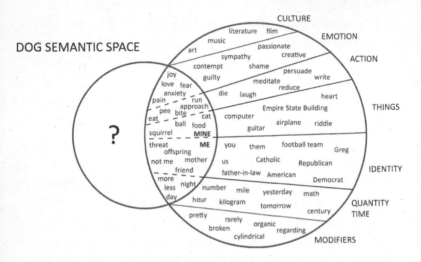

One possible representation of the semantic space of dogs and humans. The human semantic space contains a vastly greater number of concepts. A dog's semantic space, although smaller, may contain representations for which we do not have words. Only where these two spaces intersect is communication possible. "Me" and "mine" are central to both. The semantic categories for dogs are not distinct—perhaps blurring the representation of object and action, for example. It is important to note that even though I have used words to illustrate a dog's semantic space, that is for the reader's benefit. A dog might have those concepts without the words for them. From a dog's perspective, it is a nonlinguistic semantic space. (*Gregory Berns*)

of Vienna, tested eighteen dogs that had been trained to nose-poke a touchscreen. The dogs were rewarded for touching a picture of a person who was either smiling or making an angry face. Huber then tested how the dogs reacted to pictures they hadn't seen before. Critically, the test pictures only showed half the face—either the upper half with the eyes, or the lower half with the mouth. The dogs still picked out the correct emotions. Huber concluded that the dogs were able to generalize the concept of "happy" and "angry" from the local features contained in either the mouth or the eyes, and then apply that knowledge to faces they hadn't seen before.

So dogs process the information in our faces, but how was a human face represented in the dog's brain? Most animals seem to represent a face staring at them in only one way: as a threat. But not dogs. They are one of the few types of animals that have developed the ability to look back at humans without fear or aggression and even interpret something about the facial expression, including emotion.

One possibility is that through simple repetition and constant exposure to human faces, dogs build "look-up" tables in their brains. For instance, a dog could learn that when a human has an upturned mouth and squinty eyes, something good usually happens. But that would not really be a semantic representation of the facial expression for "happy." Alternatively, dogs might possess neural hardware for face processing similar to ours. If so, they would enter the world primed to process faces like human infants do, though perhaps not to the same extent.

It was my daughter Helen who prodded me to answer the question about face processing in dogs. She had been involved with the Dog Project since its inception and had to come up with an experiment for her seventh-grade science fair.

"Do you think dogs recognize their owners' faces?" she had asked me.

"I don't know," I replied. "Why don't you design an experiment to find out?"

What she came up with was simple in concept. Take pictures of the owners and show them to the dogs while we scanned their brains. As in all fMRI experiments, she would need a control condition. After further discussion, we decided that pictures of people the dogs didn't know would be the most appropriate control. And just for kicks, we decided to add pictures of familiar and unfamiliar dogs, figuring that dogs might be better at recognizing members of their own species.

We rigged up a translucent screen for the MRI. This way, we could project images for the dogs to see while the owners ducked out of view. Considering how well the dogs tolerated the scanner at this point, we hoped they were ready for this kind of screen time, even though none had shown any interest in watching screens at home.

Alas, things didn't go all that well. Many of the dogs got bored or became anxious when they couldn't see their real human. Only half the dogs stayed in the scanner long enough for us to collect sufficient data for analysis. Helen's results were inconclusive, and she took second place. But, like all good experiments, her project opened up a new line of inquiry.

Rather than giving up, I approached my colleague Danny Dilks. Danny used fMRI to understand how the human brain processed faces. Large chunks of our brains are devoted to that task. In fact, in both humans and monkeys, a swath of the temporal lobe responds preferentially to pictures of faces as compared to inanimate objects. The region is so specific to face processing that it had been named the "fusiform face area," or FFA. Danny was well versed in the FFA and the growing catalog of other face-processing regions in the brain. Helen had designed the seeds of the experiment, and with Danny's help, we altered it to match what had been shown to work in humans. Helen's experiment had focused on facial recognition, but that, it turned out, was a very complex process involving more than face processing. Danny's idea was to take a step back and determine if dogs had the equivalent of an FFA.

Instead of showing pictures of familiar and unfamiliar faces,

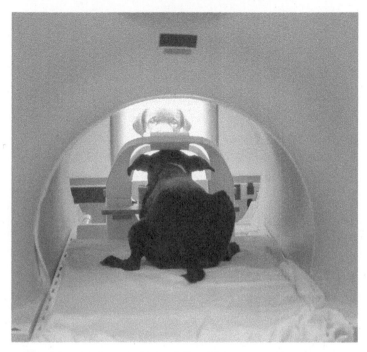

Scanning Callie while she watched pictures of faces. (*Gregory Berns*)

we showed pictures of generic human faces, dog faces, everyday objects, natural scenes, and scrambled images. A face-processing region should respond to pictures of faces but none of the other categories. It was still a difficult experiment, because the dogs had to look at two-dimensional images instead of the real thing, and only a subset of dogs was able to complete the scan sessions.

Nevertheless, the result was unambiguous. An area of the canine temporal lobe showed the characteristic signature for faces. And just to be sure, we tested the face-region with both static photos and brief movie clips. The result held, and we dubbed it the "dog face area," or DFA. Another research group confirmed the finding a year later.

The answer to the face question gets to the heart of communication. What can a dog's brain represent? Clearly, dogs come wired to process faces, and increasingly, researchers are uncovering dogs' abilities to read emotions, too. Faces and emotions are the types of things that would seem to be in the intersection of the Venn diagram of human and dog semantic space. Apparently, we both care about them.

As it turns out, dogs and primates are not the only animals that care about faces. Sheep do. Goats do. And amazingly, so do some birds. In a remarkable paper published in 2012, researchers reported on twelve wild crows that had been captured near Seattle. The twist was that the people who caught the crows wore identical masks. Once in captivity, the crows were cared for by people wearing different masks. This went on for a month. Then, the crows were administered a radioactive tracer while they were exposed to a person wearing either a captor's mask or a caregiver's mask. The tracer temporarily accumulated in the parts of the brain that were active during exposure. The crows were briefly anesthetized, and their brains scanned using positron emission tomography, which allowed the researchers to see which parts of the crows' brains had been active. The captor's mask had activated the amygdala and parts of the brainstem, suggesting that a captor's face was associated with fear and avoidance. In contrast, the caregiver's mask activated the equivalent of the caudate nucleus—the same region that in dogs was involved with positive emotions, motivation, and the desire to approach.

We should not be surprised by the ability of these animals to process faces. Indeed, it seems that many social species have it. Cows can discriminate between pictures of familiar and unfamiliar cows. Elephants, like dolphins, appear to be able to recognize their own faces in mirrors. It is not clear, however, whether cats can recognize faces. Maybe they just don't care.

The word-object experiment took far longer than we had anticipated. It was only after we made it as simple as possible that the dogs performed at a level above what they might be doing by chance. Two words, two objects, and it still took six months of training. The owners practiced at home, and every two weeks we tested the dogs' performance by giving them ten trials. The two objects were placed against a wall. When the owner spoke the name of one of the objects, the dog was supposed to go to the correct one. When the dog got eight out of ten, they were ready for scanning.

Callie didn't make the cut.

I was too focused on hammering out the details of the experiment to have the time and energy left to teach Callie to the required level of performance. But the debugging we had done was invaluable for getting the other dogs over the hurdle.

Even with the dogs who reached the 80 percent criterion, it was hard to tell what they understood. Each of them would develop idiosyncrasies during the test session. Some would favor one side, even though we alternated the locations of the objects on each trial. Others would favor a particular object. And the biases weren't consistent from session to session. Nevertheless, we had to move on to scanning. If we didn't, we would have soon had a mutiny. The owners were tired of training this task, and the slow pace of progress had diminished everyone's motivation. So we scheduled the first group of dogs for MRI testing.

Since I wasn't sure how this experiment was going to work out, it was important to design the MRI portion to contain several possible outcomes. The primary goal was to use fMRI to see how the dogs processed the two words they had been taught. This would be a tall order. Even if they understood the difference between the two words, it might be difficult to tell them apart from the brain scans. It's difficult in humans, and unlike in Gallant's experiments, the two words we had taught were semantically similar, as they both referred to objects. Perhaps if the objects differed grossly in some feature—like size or texture—we might detect a difference in the part of the brain that processed those features. But this was a longshot.

So we added a control condition used commonly in human language experiments: nonsense words. We took all of the names the owners had given the objects they had used in training their dogs and fed them into a computer program that would spit out non-words that were matched for number of syllables and bigrams. (Bigrams are two-letter sequences in words, like *sh* or *ng*.) The non-words included things like: *bobbu, prang, cloft,* and *zelve.* In addition to comparing the response to the words the dogs had been taught, we could also compare the response to these non-words.

Of course, the dogs had no way of knowing what was a real word and what was a non-word from the English language. If we could detect a difference in their brain response to the words and non-words, this would mean they could at least discriminate between something they had heard frequently and something they had never heard.

Finally, we added one more control condition. After each word was spoken, we would show the dogs an object. Most of the time, they would see the named object. But about a third of the time, we would show them a novel object. If the dogs knew what the words meant, presumably they would be surprised by the appearance of the wrong thing. And if they were surprised, this should be detectable in the brain. As a further control, the nonsense words would always be followed by novel objects.

The MRI scanning went well for most of the dogs. If anything, the humans found it more entertaining to say the nonsense words than the dogs did to listen to them. A few dogs got excited when they heard the familiar words, even to the point of running out of the scanner to look for the objects. It was unfortunate that, despite such obvious evidence of comprehension, we wouldn't get brain data from these dogs. But most of the dogs stayed put, and we ended up with data on a dozen of them.

It was a good thing we included nonsense words. Pseudowords have been used for decades in human language research, but human subjects' responses to pseudowords depends on what they are asked to do with them. In comparison to real words, pseudowords don't activate the more posterior language regions of the temporal lobe, which seem to be necessary to process the semantics of words. Because pseudowords have no meaning, no semantic processing occurs. Pseudowords tend to activate the superior temporal lobe more than real words do, however. The superior temporal lobe is a primary auditory region responsible for processing fundamental aspects of sound, like volume and pitch. It is also the region we found in dolphins' brains that receives auditory input. Increased activity in the superior temporal lobe in response to non-words is thought to be due to the novelty of the non-words. A strange word captures a person's attention and requires more mental effort to process.

Remarkably, we saw the same thing in dogs. The nonsense words activated the upper portion of their temporal lobes more strongly than the familiar words did. This proved that they could tell the difference between words they had learned and those they'd never heard before.

The results also underscored a fundamental difference between dogs and humans. Although dogs have a rudimentary capacity to differentiate meaningful words from nonsense, we did not find any evidence that they understood the familiar words as referents to objects. If they had, then we should have seen an increase in some brain region associated with recognition, perhaps the visual cortex or a different part of the auditory system. Instead, the learned words caused *less* activity. It was as if the dogs habituated to the words they knew and paid more attention to the ones they didn't.

Novelty triggers cognitive processes critical for survival. For animals, novelty could mean a new source of food, or it could mean the appearance of a new predator. Novel events require immediate action and, additionally, alter neural pathways so the animal can learn from the experience. For humans, the same processes are in play, but novelty is also a starting point for deploying symbolic and semantic processing systems. When we encounter something new, we can't help but categorize it. Our fMRI results for the dogs suggested that their language processing didn't go much deeper than novelty. At a minimum, the results showed that there wasn't anything going on that was similar to the way humans use language.

In addition to novelty, the dogs seemed to process the words in terms of actions associated with the objects. In our experiment, both objects could be nudged or picked up with the mouth. So even though we taught the dogs two words, it is possible that we didn't see any difference in semantic representation because the associated action was always the same.

An action-based semantic system would make sense for an animal. For an animal lacking language capability, there would be no need to symbolically represent the names of things. But knowing whether a thing should be picked up, chewed, or eaten would be very important.

It may be that in a dog's semantic space, actions and things are very close, which would explain why it was so difficult to teach the dogs the names of things. The semantic representation for "squirrel" might be equivalent to "chase and kill," while "ball" becomes "chase and retrieve." The fact that we eventually taught the dogs the names of two objects showed that it wasn't impossible to teach them something about the difference between the two words, but the imaging results suggested that the mechanism they used for encoding meaning was different from the mechanisms that humans use.

Humans represent the world with nouns. We name everything. There are roughly ten times as many nouns as verbs in the English language. Infants learn the names of things before they learn the names of actions, although the reason remains unclear. There was no denying that differences in language processing resulted in a different subjective experience for dogs as compared to humans, but this didn't rule out understanding what it was like to be a dog.

Quite the opposite.

The fact that we can discover these differences and interpret them shows that we can understand what it's like to be a dog. As we had seen throughout our studies of animal brains, it might simply require a shift in perspective—in this case, from a noun-based worldview to one based in action.

If the semantic space of dogs is organized around actions rather than objects, then this would explain why they failed the usual tests of self-awareness, namely, the mirror test. Humans know that a reflection is a visual representation of something or someone. We take it for granted that the reflection is not the thing itself. But this cognitive operation requires the mental hardware for symbolic processing of things. If dogs' brains are not wired to symbolically represent things, then they do not have the ability to link their reflections with a sense of self.

This would not mean that a dog doesn't have a sense of self. It would just mean that a dog doesn't have the ability to represent that self abstractly, either by name or visual image. My beloved Callie probably didn't have abstract representations of me or my wife or my children. No, I was just *that guy who feeds me hot dogs in the loud tube, and so I interact with him in a certain way.* My wife was *that other person who feeds me and pets me but doesn't play with me, and with whom I interact in a different way.* Callie's mental representations might be defined entirely by these interactions. We might say the representations are transactional.

In an action-based worldview, everything would be transactional. Even emotions might be represented as actions. Fear would become *that feeling in which I need to get away from something.* Loneliness would be *that feeling which is lessened by waiting by the door until it opens and then goes away.*

I am not just anthropomorphizing. The words I used were a necessary construct of communicating an idea in written form. A dog could not think the literal words I wrote, because the dog doesn't have the brain architecture for thinking in words. An action-based semantic system does not mean, however, that fear is just the set of motor programs that an animal implements to escape something unpleasant. The motoric aspects are important, but so is the subjective awareness of what's happening, for that is where we have common ground.

With perhaps the exception of a few apes and dolphins, I suspected that a semantic bias toward action was characteristic of all animals. If so, then I needed to reconsider my attempts to communicate with them. Animals do not have the capacity to take our point of view, but we have the capacity to take theirs. What if our emphasis shifted toward communication of actions instead of communication of names of persons and things? Then, perhaps we could come to a better understanding of what it is like to be a dog, or a bat, or a dolphin.

Maybe we would even learn what they had to say.

Unfortunately, many of the Earth's animals might be gone before we figure out how to communicate with them. For them, someone would have to speak for the dead.

Chapter 9

A Death in Tasmania

The tiger paced the perimeter of his enclosure. He hadn't seen his usual caretaker in months, a span of time for which he had no concept, but the tiger knew time had passed. It had been the middle of winter when the human female had last fed him and opened the door to his sleeping den. Other zookeepers had come and gone since. But now the days were growing longer.

The door to his den was closed to keep him outside and on display for visitors, just as was done in every carnivore's enclosure. The sun was bright, and he had no choice but to continue pacing the area in search of relief. Without the shade of the eucalyptus tree that once loomed over the corner of his enclosure, he had no place to rest. Perhaps a human would take pity on him and beckon for his den to be opened. But humans had become scarce. The few who still came spent their time at the other end of the zoo, regaling the true big cats rather than looking at a dingy, smelly creature like him.

For the "tiger" was not really a tiger, at least not in the sense that scientists thought of them. He was a thylacine. A marsupial.

The tiger moniker had been coined a century earlier by settlers and convicts from Britain who thought that any animal with stripes upon his back was of the same family of animals.

If any visitors had ventured to the rear of the zoo, they would have seen a ragtag mob of kangaroos and a few deer milling about their paddocks across from the thylacine's. But kangaroos and deer were not the main attraction of a zoo in Tasmania. Any bushwalker knew them well, and the thylacine wasn't terribly novel to the public either. Days passed without anyone visiting the animals of Tasmania. And now his coat was a matted mess with patches of skin showing. His tail dragged behind him without vigor. Had anyone wandered by, they would have been hard-pressed to see his famous stripes.

Without a cloud in sight, the sun baked the hardpan of his enclosure to the point that it was uncomfortable to walk on. His best strategy was to lie down in the midday sun and let the soil cool beneath him. Basking went against his natural instincts. In ordinary circumstances the thylacine was crepuscular, preferring the vagaries of dusk and dawn when the snow gums cast their shadows with inky fingers across the bush. Only at those times did the full benefit of his stripes become apparent, rendering him all but invisible to the hapless wallabies and wombats.

It had been three years since he had killed a wallaby. Three years since he had seen another one of his kind.

Across town, Alison Reid was preparing lunch for her mother. The thirty-one-year-old with pale white skin and chestnut brown hair could sense it was going to be a warm day. Although Alison had awoken even before the sun rose, her mother had only just gotten out of bed. Having been evicted from the curator's home at the Hobart Zoo months before, in June 1936, she and her mother were staying with relatives. Dad had been dead two years and Mum had never been the same. Her mother had taken to bed with increasing frequency. As Alison prepared the midday meal, her thoughts turned to more pleasant times.

As the daughter of the zoo's first curator, Alison had grown up surrounded by animals. Her father had also taught her how to preserve the remains of animals that had died so people could still see them and learn about them. She still remembered the secret recipe for the solution of alum they soaked the hides in to make them supple and last for decades. Alison had learned the artistry of constructing a mount to drape the preserved skin over, bringing the dead back to life. She had mounted cats and dogs for owners who couldn't bear to spend a day without their deceased pets. Alison was so good at her craft that the Tasmanian Museum had hired her as a taxidermist at the age of seventeen.

But Alison's real love was for the living animals, and she had spent most of her spare time at the zoo. It seemed ages had passed since she had helped her father nurse Sandy and Susie, a pair of lion cubs, or seen her favorite feline, Mike, a leopard with whom she had been famously photographed for the newspaper. Alison still had the clipping: "Beauty and the Beast at the Hobart Zoo: Girl Whose Greatest Chum Is a Full-Grown Leopard." The picture with Mike that they had published was not her favorite. She had been trying to smile while wrestling him into her lap. The one-hundred-pound adolescent was being stubborn that day, and in the photo he was half-in half-off her lap. But still, it had made her a celebrity! The article told everyone how she had taken Mike for walks along the River Derwent when he was a cub. But now the article only stirred a deep ache for how things once had been. How she longed to be walking Mike along the bay, with the sun setting behind Mount Wellington.

Alison was practically crepuscular, too, her favorite times of the day being dusk and dawn. That was when she used to feed the animals and tend to their needs.

She thought of the thylacines. She had never known when a trapper would show up with a tiger, usually with its leg mangled from a snare. Alison and her father had always taken them in, and they had done the best they could to bring them back to health.

The thylacines were nervous creatures, and they gave off an unpleasant odor when agitated. It took months for them to get comfortable around her. Eventually, they would learn the routine of the zoo: out in the morning, fed in the afternoon, and back in their den at night. And they loved to eat. Despite their shyness, if anyone was late for their afternoon feeding of rabbit and calf carcasses, the tigers would make their hunger known. A hungry tiger agitated for its food with a coughing bark, sounding like a consumptive hacking up phlegm.

It was true that thylacines were cranky creatures, even hard to love. But that didn't justify the disgust most people had toward them, or how they had been labeled stupid marsupials. Alison sympathized with the thylacine. For she, too, had experienced the slap of close-minded men in powerful positions. She was the most qualified person in Hobart to run the zoo. And here she was, making tea instead of feeding the animals.

Alison wondered if anyone would remember to feed the thylacine.

The midday warmth did not last long. It was early September, and by four o'clock, the sun was edging toward Mount Wellington. All of the inhabitants of Hobart, man and beast, would soon be in shadow.

A photon, birthed in the sun eight minutes earlier, careened off dust in the atmosphere above the Indian Ocean and headed toward Sandy Bay. It bounced off the door of the thylacine's den, its final trajectory taking it through the delicate membranes of the tiger's eyelid and slamming into his retina. There, in a minor chemical explosion, it gave up its remaining energy and sent a signal into the animal's brain.

Wake up.

The tiger blinked open his eyes. Thankfully, it was no longer bright, and his pupils dilated accordingly. He scanned his surroundings for movement. A hundred million years of evolution had given him a visual system perfectly adapted for detecting things that moved along the horizon. But all he saw now were the kangaroos hopping about. Their scent aroused his hunger. The tiger had watched and smelled the kangaroos for a long time. He knew there was no point in expending effort to reach them.

With no apparent prey within his reach, he coughed. In the past, this had resulted in the appearance of the human female. She would toss him some dead animal. If he was lucky, it would be fresh and the blood would flow freely as he ripped open the chest and abdomen. Usually, though, it was just a piece of meat, dead for days.

The hunger was bad. Did he eat yesterday? He couldn't remember. That didn't mean he hadn't eaten. He just couldn't remember. There was no shit in his pen. Probably not.

He coughed some more. He could hear people in the distance, but his keen nose couldn't smell any meat, and so, after a few more attempts to raise attention, he lay down.

The temperature fell fast. The tiger slept.

He was home.

Ever since leaving his mother's pouch, he had lived his whole life here. The predawn mist in the Florentine Valley had not yet begun to burn off. The twilight was cool but not uncomfortable.

From his vantage point on the ridge, his sharp eyes scanned the tops of the giant eucalyptus trees poking through the fog. The vigor of youth flowed through him. He yawned open his massive jaws, inhaling deeply to sample the scents floating up from the valley. They were all there: wallaby, wombat, devil.

He padded down a slope, slipping without a sound into the forest. The other animals of the twilight would either be waking up or returning from their nightly journeys. He was in no hurry. His kind were deliberate in their hunting. There was no need to chase down prey when, cloaked by stripes, he could sneak up on them.

As he glided through the undergrowth, he caught a flicker of motion in a fern. There was no wind so it could only be another animal. Another thylacine? Not likely. He hadn't seen one in weeks.

He continued on in his original direction, making a mental note of where the fern was. With each step, he made a small correction to one side, which had the effect of turning his straight-line path into an arc. The arc grew tighter and turned into a spiral. Now, he was behind and could see the tiny wallaby—a pademelon—foraging in the low grass beneath a myrtle.

The thylacine sprang upon the wallaby and had his neck in his jaws before they hit the ground. There was no struggle. The thylacine opened up the belly and ate the organs.

Ever confident, he strode off, leaving the carcass for the devils.

The last known Tasmanian tiger, or thylacine, after its scientific name *Thylacinus cynocephalus*, died on September 7, 1936. His death was recorded simply a week later by the Hobart City Council: "The Superintendent of Reserves reported that the Tasmanian tiger died on Monday evening last, 7th instant, and the body has been forwarded to the Museum." In later years, people started referring to him as Benjamin, but while he was alive he had no name.

At the time of Benjamin's death, nobody knew that he would be the last thylacine to be definitively observed. According to international conservation standards, if an animal is not seen for fifty years, it is considered extinct. And so, in 1986, the thylacine was moved from the endangered column to the extinct category. But Tasmania remains one of the last great wildernesses in the world, and many people believe thylacines still exist in the remote bush.

For those who wish to see a thylacine today, there exists only three minutes' worth of grainy black-and-white film of captive thylacines. When I saw this silent footage, almost a century old, it spoke to me. In appearance, they look much more like dogs than tigers. Coming from the lineage of marsupials, however, they were separated from dogs by 100 million years of evolution. Clearly their appearance is a case of convergence, but it made me wonder: If thylacines look like dogs, did they think and behave like dogs, too?

I became obsessed with the question of the thylacine's mind. This was a potentially dark road. Pursuing cryptids—supposed creatures like Bigfoot and the Loch Ness monster, or likely extinct ones like the thylacines—was not generally considered worthy of academics. And yet, the thylacine called to me.

Of course, I am a sucker for anything remotely canid, and the thylacine hit all the right notes. Its scientific name means "pouched animal with a dog-head." To make matters worse, the thylacine had become a tragic figure. Not much is known about the origins of the thylacine, but there's no mystery about why they disappeared.

When the crown mammals appeared roughly 150 million years ago (mya), they still laid eggs like the dinosaurs. Only a few descendants of these egg-laying mammals exist today, and they are called monotremes. The platypus is the best known, but among the others are echidnas, which are hedgehog-like creatures with long beaks that feed on insects. Monotremes are found exclusively in Australia and New Guinea.

As the great continent of Gondwana broke up, the crown mammals separated onto the subcontinents and began their own evolutionary trajectories. By 120 mya, some of these mammals had evolved the mechanisms to gestate their eggs internally. The eggs didn't gestate very long, and the young were born alive but small. They continued to develop in an external pouch that contained multiple nipples. These were the first metatheria. Their descendants today are known commonly as the marsupials, of which the thylacine was one. The final split in mammalian evolution occurred 100 mya with the appearance of the placental mammals. The placentals gestated their young internally far longer than the marsupials did and eventually beat the marsupials in the Darwinian competition for resources. Only in Australia and New Guinea did the marsupials maintain dominance.

Australia had the benefit and curse of isolation. Australian species had the benefit of evolving without competition from plants and animals in the rest of the world. But that also made them more susceptible to suffering from competition when it finally arrived.

Thylacines used to live throughout Australia, and as the apex carnivorous predators, they had no competition for food for thousands of years—until the humans showed up. The first Aboriginal people arrived 20,000 years ago, and the rock art they left shows multiple depictions of thylacines. Thylacines co-existed with humans for a time on the mainland. In 1966, a mummified thylacine was discovered in remote southwestern Australia. Radiocarbon dating put its age at 4,650 years old. But whether it was due to competition with humans directly, or with their dogs, the thylacines were no match for the newcomers, and the mainland population disappeared, leaving only a few thousand thylacines behind, all trapped on the island of Tasmania, which, after the most recent glacial retreat, had been cut off from the mainland.

It was really just a matter of time.

By the time I found out about the thylacines, their minds seemed lost forever. They had been extinguished just as scientists had begun looking seriously at animal behavior. The few captive thylacines that remained in the end kept mostly to themselves. Zookeepers noted that they tended to bask in the sun, although the trappers thought the thylacines had been nocturnal. Nothing was known about their social lives.

My obsession with the thylacine continued to grow, and I dug into everything I could find about these sad-looking creatures. As I soon found, though, I was not alone. A small but dedicated international community had coalesced around the thylacine. Some people were driven by the quest for knowledge about the animal. Some were motivated by the realization that countless other animals were on the path to extinction, and that maybe something could be learned from the thylacine's story. And many believed that the tigers still existed in the remote wilderness of Tasmania. In fact, tiger sightings are claimed with regularity, and there are even a few videos posted on YouTube showing fleeting images of animals people claim were tigers. Of course, none of these are convincing, and no reputable scientist takes them seriously. In 1984, Ted Turner reportedly offered $100,000 for proof that the thylacine still existed. The prize has gone unclaimed.

One of the few scientists who does take the thylacine business seriously is Michael Archer. A respected paleontologist, Archer is a professor at the University of New South Wales in Sydney and was formerly the curator of mammals at the Queensland Museum. Archer has received many awards for his work in mammalian evolution. And he has tried to clone the thylacine.

Seriously. Archer was ahead of the times in trying to de-extinct the thylacine from specimens in the museum collection. That was in 1999. The project was ultimately abandoned for lack of high-quality DNA, but if anyone could help with what I was thinking, Archer could.

I was going to speak for the dead in the only way I knew

how: through the thylacine's brain.

With the success of the dolphin project, we had shown that you could learn something useful about an animal's perceptual world by looking at its brain. The dolphin brains had been over a decade old, and it had been challenging to squeeze enough signal out of them. But we did. A thylacine brain, if one still existed, would be nearly a century old, at best. Maybe we could do the same for the thylacine.

Archer responded right away. He didn't know the state of thylacine brains, but he knew who did.

Stephen Sleightholme has been working since 2005 to collate a database of all known thylacine specimens. He calls it the International Thylacine Specimen Database (ITSD), and by 2013, the database was so large that it barely fit on a DVD. I had never seen anything like it. And like Archer, Sleightholme responded immediately.

According to Sleightholme, there were four known intact thylacine brains: one each in Australia, Germany, England, and the United States. Two of the specimens were believed to be partially damaged. The one in the United States was held by the Smithsonian Institution.

The Smithsonian's collection of specimens runs far deeper than what is on display at the museum in Washington, DC. Darrin Lunde and Esther Langan, in the division of mammals, knew all about the thylacine brain. As one of only four in the world, the brain was a national treasure and therefore not on public display. Plus, who besides cryptozoologists cared about it?

I cared about it. A lot. That pickled brain in the Smithsonian's vault might contain the keys to the thylacine's mind.

To prove that it was real, Lunde sent me a picture of

the brain in a jar of formaldehyde and a copy of the original cataloging.

The card catalog was for the Division of Physical Anthropology and was clearly not designed to record animals, but the card contained crucial information. The specimen had been received on January 11, 1905, which was also the date of death. Sex: male; age: adult; occupation: m N.Z.P. Presumably this was shorthand for "member" of the National Zoological Park, aka the National Zoo.

According to Sleightholme's database, the former owner of this brain had been caught as a pup along with his mother, sister, and one other sibling in 1902. The family was sold to the Launceston City Park Zoo in Tasmania, which then sold the family to the National Zoo in DC. Much is known about the story of this first thylacine family. It isn't pretty.

Actually only the mother had been captured. It wasn't until she arrived in DC after a grueling three-week ocean voyage and train ride across the United States that the zookeepers noticed she had young in her pouch. That any of them survived was a miracle. One of the pups died after nine days at the zoo.

Sadly, the mother had been so weakened by the trip that she died after four months from an "intense inflammation of the intestinal tract" and an extensive load of tapeworms throughout her body. That left two surviving pups.

The male lived until January 11, 1905, when he, too, died of hemorrhagic enteritis. His head ended up in the hands of an anatomist, Aleš Hrdlička, who extracted the brain—the one that now sat in the Smithsonian's vault. The female lived alone for another few years.

What was learned from this great opportunity to observe

The two surviving thylacines at the National Zoo. Image originally published in the *Smithsonian Institution Annual Report*, 1903, p. 66. (Smithsonian Institution Archives Image #NZP 139)

thylacines? Apparently, nothing. No scientific study of these creatures was ever undertaken while they were alive. And so I faced the prospect of a forensic reconstruction of the thylacine's mind from the only remaining artifact: its brain.

At the Smithsonian, Lunde and Langan told me that the thylacine brain had been subjected to an MRI scan many years ago, but the results were of such poor quality that they didn't think it was worth publishing. It was a longshot, but I asked if they would be willing to try again, this time with a more powerful scanner and new sequences designed for postmortem brains.

They were up for it. But because of the rarity of the

specimen, we would need to follow a rigid protocol for handling and protecting the brain.

As far as I could tell, nobody had ever attempted to scan a brain this old. The thylacine we were scanning had died in 1905, making the specimen 110 years old. Even though it had been in formaldehyde ever since, there was no way to know what a century of pickling had done to it. As we had learned with dolphin brains, even a decade in preservative changed a tissue's properties. But it would be dangerous to extrapolate from ten years to a hundred. Maybe the changes got worse, maybe they plateaued. We would need to spend a fair amount of time experimenting with different sequence parameters to squeeze as much signal out of the thylacine brain as we could, assuming there was anything left to give off a signal.

We would be under time pressure, too. The Smithsonian's generosity in lending out the specimen was not unlimited. It would have to be kept in a secure location, and I wouldn't dare advertise the fact that we had a national treasure. The ideal scenario would be to scan it and send it right back. The less time I was responsible for it, the better.

The more we could prepare for the thylacine in advance of getting the brain, the smoother scanning would go. So that we could work out all the scanning details, Lunde offered to send the brain of a raccoon from the same era. That the Smithsonian continued to hold the brain of a raccoon from the last century was entirely of historical interest, not scientific. Crucially for us, the raccoon brain had been extracted and preserved by the same anatomist who had worked on the thylacine. So his preservative recipe would likely be the same in both cases.

A week later, a wooden crate the size of a shoebox arrived in the lab. It was against air transportation regulations to ship flammable liquids, so the specimen had been wrapped in moistened gauze and double-bagged in sealed plastic. It was smaller than I expected, about the size of a walnut, but then again, I was used to the brains of large carnivores.

I sandwiched the brain between two pieces of sponge and jammed them into a cylindrical plastic container. I topped it off with an inert liquid that would not give off any magnetic resonance signal.

The brain was too small to put in the head coil. The pickup sensors would be too far from the specimen to pick up much signal, and I didn't have the time or money to build a miniature coil. The next best thing was a flex coil. This was standard equipment with the MRI scanner. A flex coil contained two pickup elements embedded in a foam sheet that could be wrapped around any body part. In clinical imaging, flex coils were used in areas where it would be awkward or impossible to get the part of the body in a cylindrical coil, such as the shoulder.

It wasn't ideal, but I wrapped the flex coil around the specimen container and sent it into the center of the MRI. The MRI gradients would likely limit us to a resolution of about 1 mm. The limit on resolution would mean that each voxel in a small brain would contain more structures than a voxel in a large brain. I hoped the resolution of the raccoon brain would be sufficient for us to see the structure inside it.

Peter and I set up the structural sequences and hit *Scan*. The MRI went through its preparatory clicks and buzzes and launched into the scan with the sound of a submarine about to dive. The specimen was so small that it was all over in two minutes.

The images popped onto the screen. They looked good. Really good. We had pushed the gradients to their limits and achieved a resolution of 0.3 mm, better than I had expected.

We could see everything: caudate, corpus callosum, cerebellum, hippocampus. Crucially, the images had nice contrast between the gray and white matter. This meant that the tissues hadn't degraded into a uniform mush after a century in formaldehyde.

DTI was tricky. Using the parameters that we had used in the sea lions and dolphins only produced faint images of mostly pixilated noise. I consulted with Karla Miller, the MRI physicist back in Oxford. She told me to calculate how fast the preserved brain was responding to the magnetic fields and adjust the scan times to capture the signals before they faded away.

There are two quantities that describe how something behaves in a magnetic field. T1 refers to the time it takes for a tissue placed in the magnet to become fully magnetized. T2 is for the time it takes for the protons to get out of sync after being hit with a resonant radio wave. Together, T1 and T2 are called relaxation parameters, and each tissue has different ones. These differences allow for the generation of contrast in the images. A healthy brain, placed in a 3 Tesla magnet, has a T1 of 1,300 milliseconds (ms) for gray matter and 830 ms for white matter. T2 is even faster, about 80 ms, which means that the MR signal in a healthy brain decays very quickly. The preserved dolphin brains, which had been in formaldehyde for ten to fifteen years, had a shortened T1 of 350 ms. In the raccoon the T1 had fallen to 200 ms, and the T2 was a brief 30 ms.

The shortened relaxation times meant that the MR signal decayed away quickly. We would have to scan faster if we were to catch these fleeting emissions. But everything in MRI was a tradeoff. Faster scanning meant driving the gradients harder. We had a choice: scan faster or scan with more intense fields. But we couldn't have both.

I spent hours at the MRI console. Through trial and error,

I would adjust the scan speed (called TR for time of repetition), and then try to find the maximum gradient power that the system would tolerate. Sometimes the scans would abort immediately with a warning that the gradient power had been exceeded. Sometimes, the scans would start, but abort after an hour or two of scanning. It was frustrating.

After a week of fiddling, I thought we had hit the sweet spot of squeezing signals out of the brain without frying the scanner. The images were still noisy, but if we repeated the scans multiple times, we could average them together and reduce the effects of noise.

I emailed Lunde and Langan to say we were ready for the thylacine.

Although much has been written about the thylacine, only a few people can be considered experts. Archer, a paleontologist, and Sleightholme, the curator of the specimen database, are two. A third, Cameron Campbell, maintains a remarkable website called The Thylacine Museum that is a comprehensive source of information about all things related to the thylacine. Information on everything from its history to its anatomy to the ongoing debate about whether they still exist in the wild can be found on Campbell's site.

As I sifted through Sleightholme's and Campbell's databases, it became apparent to me that much of the information had been sourced from a few individuals, most of whom were dead. The true father of thylacine research was Eric Guiler, an Irish marine biologist who moved to Tasmania in 1947. Guiler always considered the possibility that thylacines had not gone extinct. His expeditions and interviews with hunters and trappers who had actually known the thylacine provided the most detailed history of this remarkable animal. Guiler died in 2008—six years after suffering a stroke while on a thylacine expedition.

Guiler thought that the thylacine's decline on the mainland, perhaps accelerated by humans, was more likely due to climate change. Until 6,000 years ago, the mainland experienced a wetter climate, which would have fueled the growth of plants that animals such as the wallaby would have fed on. The thylacines preyed on these herbivores, while forests would have provided cover for them when hunting. But by 5,000 years ago, the mainland climate had become much drier, closer to what it is today. The thylacine's habitat shrank, and soon they disappeared.

The land bridge to Tasmania had flooded at the end of the last ice age, leaving the island's population of animals cut off from the mainland. Being farther south, toward Antarctica, Tasmania's climate was considerably more temperate and varied than that of mainland Australia. Tasmania's habitats ranged from rain forests to scrub highlands and everything in between. Perfect for the thylacine.

As far as anyone knows, the Tasmanian thylacine population had been doing okay until the arrival of English settlers. Named for the Dutch explorer Abel Tasman, who discovered the island in 1642, Tasmania had become an infamous stopping point for those traveling to the southern Pacific. As part of Captain James Cook's exploration of the region in 1773, the HMS *Adventure* anchored in a well-sheltered bay on the southeast shore, which became known as Adventure Bay. William Bligh had a layover there on the HMS *Bounty* in 1789 on his way to Tahiti.

The explorers did not log any sightings of the thylacine.

They probably didn't venture ashore far enough to see them. Even so, Adventure Bay acquired a reputation for its abundance of fresh water and its plant and animal life. The first reference to the thylacine dates to 1805 with reports of a "tyger."

The early settlers brought sheep for food and wool. Were it not for these poor sheep, it is possible that the thylacines would have survived. But the sheep were easy prey for all the island's predators: dogs, devils, and thylacines. It became such a problem that a bounty was placed on all three. Thylacines, which settlers referred to as hyenas, brought the largest bounty—twice as much as for devils and wild dogs. To make matters worse, the bounty escalated for every twenty thylacines killed. It was a clear incentive to rid the island of them.

Landholders complained that the thylacines were killing their sheep through the late 1800s. The number of sheep killed was likely inflated, and wild dogs probably accounted for the majority of the carnage. Nevertheless, the thylacine took the blame. Myths even began to circulate that thylacines drove sheep off cliffs.

The thylacine was not the only native animal to suffer. The other predators went into rapid decline, too, especially the native "cats," which, like the thylacine, were actually marsupials (they are now called quolls). Although the bounty scheme played a significant role, equally important was the loss of habitat. Many of the forests were cleared to make room for cattle and sheep grazing.

By the 1930s, thylacines had become increasingly rare, and conservationists thought they might be going extinct. When Benjamin died in 1936, the situation had become one of too little, too late. Calls for the establishment of sanctuaries went unheeded. Part of the problem was ignorance, and denial that the thylacine might actually be gone. The Tasmanian wilderness was still vast, and many people assumed that they were still out there. After all, people continued to report seeing them.

In 1937, a newly formed Tasmanian animal protection board dispatched a party led by a police trooper, Arthur Fleming, who was an experienced bushman, to see if they could find the thylacine. On his first expedition, he headed to the western mountains. This was extraordinarily rough terrain. Although he didn't see any thylacines, he did find tracks. Encouraged by these discoveries, he returned in 1938 with a larger team. Again, they found footprints, but didn't see a live animal.

Fleming tried yet again in 1945. For six months, his team trekked back and forth between the Jane River and Lake St. Clair. They tried everything, from setting up live sheep decoys to dragging entrails through the bush, to lure thylacines to their traps. Although they succeeded in catching just about every other indigenous animal, no thylacines were spotted. By April 1946, they had given up.

The thylacine trail went cold for the next decade. Then, in September 1957, some sheep were found mutilated in a town just north of Hobart. Guiler, who had been living there for several years, swung into action. He wrote: "The sheep had all been killed by having the throat eaten out very cleanly, there was no blood on or near the carcase and it had probably been lapped up by the killer. The nasal bones also had been eaten away and there was no wool torn off the victim."

To Guiler, the condition of the carcass was not characteristic of how dogs killed sheep, especially because there were no bites to the legs. Plus, he found footprints that could have been made by a thylacine. Expanding his search, he found that other farmers had experienced similar sheep kills over the past few months. The old trappers told Guiler that these had been typical of thylacines. One farmer even reported seeing a thylacine stealing off with a loaf of bread.

Guiler built a trap and placed it on the property where the thylacine had been spotted. But even after a year, Guiler had caught nothing.

From 1957 until he had his stroke in 2002, Guiler organized over a dozen expeditions in search of evidence that the thylacine still existed. He favored the Woolnorth region, where the sheep were killed, but several attempts failed to turn up anything definitive. To cover more territory, he eventually switched to setting snares, as opposed to cumbersome traps that were difficult to lug through the bush, and then, finally, cameras.

Guiler remained adamant that Benjamin had not been the last thylacine. Some had certainly existed in the bush for some time. But he couldn't say for how long. As the decades wore on, the technology got better and better. Yet the searches became more difficult, producing less and less. Without definitive evidence, Guiler eventually came to accept that the thylacine most likely had slipped away forever.

The thylacine brain arrived in a wooden crate, but this one was much bigger than what the raccoon came in and could have easily accommodated a human brain. The Smithsonian's card catalog said that the brain had weighed 43 grams when it had been extracted. A human brain weighs about 1,300 grams.

Peter stared at the crate and said what we were both wondering. "Why the big box?"

There was only one way to find out.

We located a screwdriver and backed out the dozen screws

securing the lid. Inside, we just saw foam packing material. I plunged my hand in, feeling for something, until I found the sealed plastic bag. My hands trembled as I pulled out the national treasure. Like the raccoon brain, whatever was inside the bag was wrapped in gauze.

We were under strict instructions to keep the specimen in a solution of formaldehyde and ethanol. Peter had a Tupperware container at the ready. I took a deep breath, tried to steady my hand, and cut open the bag. The smell of formaldehyde and alcohol was intoxicating. With the utmost care, I unfolded the gauze.

It was smaller than I expected. The thylacine had been the size of a medium dog. Dog brains are about the size of a large lemon. But what I held in my hand looked like a walnut. And it was almost as hard. Even brains fixed in formaldehyde still had some give. But not this one.

Peter was thinking the same thing and just said, "Huh."

"Let's weigh it," I said.

Peter put the specimen on a digital scale and read, "Sixteen grams."

"That's a third of its original weight," I said. "How can that be?"

"It must have shrunk."

Indeed. Concerned about the shrinkage, I ran a quick calculation. If the brain shrank at the rate of 1 percent a year, after 110 years, it would, in fact, be precisely 33 percent of its original weight. This was unexpected. I had no idea whether the brain had shrunk uniformly or the shrinkage had proceeded unevenly and resulted in some weird contortion.

It didn't look like any other brain I had ever seen. The cerebellum was a knobby collection of ridges and looked like a piece

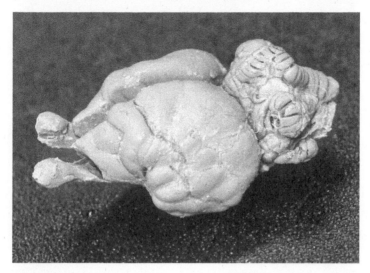

The thylacine's brain. (Smithsonian Institution, specimen USNM 125345. *Photo: Gregory Berns*)

of cauliflower. The cortex wasn't entirely smooth, and I could make out some gyrification. This was good. It meant that the thylacine had evolved a brain complex enough to require folding. And then there were the big olfactory bulbs, jutting out like a pair of antennae. Relative to the size of the brain, these were even larger than in the dog. I could see that the olfactory bulbs were attached to a piece of cortex that looked different from the rest of the brain. This was the piriform cortex—the part of the brain devoted to odor processing.

But what else could we learn about the thylacine? Size wasn't everything. I hoped that we could map the connections of its brain and figure out something more about its mental life. Were thylacines social? Did they have a big frontal lobe for problem solving?

Before Peter and I could get to the DTI scans and map the

connectivity, we needed to get a detailed picture of the brain's internal layout. We followed the usual protocol and set up the brain for a structural scan.

Peter said, "That does not look anything like a dog's brain."

The olfactory bulbs were big and projected way out in front of the brain. I could identify some major landmarks, like the thalamus, and the weird-looking cerebellum. But there was no corpus callosum. All we could make out was a thin tendril of fibers between the two hippocampi.

I emailed Sleightholme, and he told me that the lack of a corpus callosum is a hallmark of the marsupial brain. Only placental mammals have one. Instead, the hemispheres in a marsupial brain communicate with each other through a bundle of fibers called the anterior commissure. These are located toward the front of the brain and beneath my favorite structure, the caudate nucleus.

How the thylacine's hemispheres communicated was more than an academic question. As we saw in the dolphin brains, the physical layout would provide us a roadmap for how information flowed, and that would tell us something about function. To go deeper, we would need DTI data.

We started with the parameters from the raccoon brain. The first DTI images, though, were barely visible. I could make out a ghostly image of the thylacine's brain, but it was so faint that there was no way the DTI data would be usable. The T1 had shortened even farther than in the raccoon. For the thylacine, the T1 measured a scant 150 ms, almost one-tenth that of a heathy brain. I would have to push the gradients to scan faster so that we could capture these faint signals before they decayed away.

After a day of fiddling, I was satisfied that we had achieved a reasonable tradeoff between the need to scan fast and the need to apply the strongest gradient field possible. The signals would still be weak, so Peter and I programmed the scanner to collect twelve repetitions of each image. Since we were collecting diffusion information in fifty-two directions, this meant over six hundred images in total. Because of its rarity, I didn't dare leave the specimen unattended, so I babysat it all day. By 9:00 p.m., the sequence was complete.

Over the next several months, I puzzled over the scans. There was no precedent for scanning brains this old, and certainly we had been the first to try and put together a connectivity map from DTI data. The thylacine's brain was just too different from what I was accustomed to. Whether this was because it was a thylacine or because it was a marsupial, I couldn't tell. I needed help from someone who knew marsupial brains.

Hardly anybody studied marsupial brains. Ken Ashwell, a neuroanatomist at the University of New South Wales in Sydney, was the exception. He had edited the definitive textbook on the topic. I wrote to him and asked if he would like to collaborate on the project.

Ken was intrigued. He had been pursuing similar MRI approaches with monotremes, like the echidna and platypus. He agreed to look at the scans and take a stab at locating key landmarks, especially nuclei in the thalamus.

Even for the world's expert, this was not a straightforward task. Based on his knowledge of marsupial brain anatomy, Ken made his best guess at the location of structures like the medial geniculate nucleus—the landing point for auditory information, which was the same structure Peter and I had used in the dolphin's brain.

Over several months, Ken and I went back and forth. I would send structural images from the thylacine, he would locate thalamic nuclei, and, using these locations, I would use probabilistic tract tracing to determine how they were connected to the cortex. It was slow going, and I could work only intermittently. And for every new result, we had to ask a basic question: Was this specific to the thylacine, or was it characteristic of marsupial brains in general, or perhaps more specifically, carnivorous marsupials?

When I began the search for the thylacine, Sleightholme had told me that there were four intact brains in the world. The ones in Berlin and Oxford were likely too damaged to bother with. The Smithsonian specimen, although shrunken, was in good shape. That left the one in Australia.

Australians are highly protective of thylacine material. Although supposed pelts occasionally surface in auctions, they often turn out to be fakes. Most thylacine material is still in Australia, and the museum specimens are not likely to ever leave the country. For this reason, I hadn't given any thought to scanning the Australian brain.

But Ken had other ideas. He had been working with the Australian Museum in Sydney to scan some of the brains in the museum's mammal collection. Ken was interested in the monotremes, but on one of his museum visits he saw the thylacine brain suspended in a jar of formaldehyde.

In an email he wrote, "It looks to be in good condition."

Since there was no way the museum would let me borrow the brain, it was obvious that I would have to go to Sydney to scan it. Luckily, Ken's university had a 9.4 Tesla MRI for animal scanning. This was three times the strength of our machine in Atlanta. The higher magnetic field meant a stronger signal, which would let us scan at finer resolution.

The museum was willing to lend the brain to Ken, but only under strict conditions. Ideally, we would have liked to have it for a few days. It would take some time to prepare the specimen in a container of inert fluid and fiddle with the scanner settings. Every brain was different, especially these old ones. But the museum would only let us keep the brain overnight if we posted a guard. The cost was prohibitive. Plus, we would have had to buy insurance for the brain, which was no small feat, since its value was indeterminate. It would be like putting a value on the *Mona Lisa*. We compromised and agreed to borrow the brain for one day. Somehow, we would have to accomplish everything in eight hours.

So I would go to Australia. And from there, it would be a short flight to Tasmania.

Chapter 10

Lonesome Tiger

My last known location would have been registered with the nearest Tasmanian cell tower fifteen kilometers to the southeast, just before I had turned off the main throughway onto the dirt logging road. After driving for half an hour without seeing a soul, I had come to two tire tracks, marked simply as No. 5, that split off the logging road and disappeared into the forest. I had maneuvered the rental car carefully down the tracks, barely clearing the trees that had tipped near horizontal. Ignoring every indication that the forest did not want me there, I had pressed on for another kilometer until reaching a gate that blocked further passage.

Now, proceeding on foot, I almost missed the entrance to the trail, which was little more than a parting in the thicket. A small piece of tape tied to a branch was the only sign it was the right place.

I really had no business bushwalking alone. My daypack contained one bottle of water, two granola bars, and written instructions of how to find the location of the last known thylacine. My boots were soaked within the first hundred meters of entering this primordial jungle. Thirty years ago, I wouldn't have given a second thought to hiking a remote trail in Tasmania alone. But now, I realized how fundamentally stupid this quest was. Actually, the only stupid part of this adventure was that I had gone alone. Other than that, it made perfect sense.

As Peter and I had seen with other animals, interpreting brain structure required knowledge of an animal's environmental niche and its behavior. With the dolphins' brains, we had drawn upon a wealth of knowledge of their physiology and the marine environment they lived in to understand how the auditory pathways in their brains were responsible for both conventional hearing and echolocation. With the sea lions' brains, we had relied on Frances Gulland's work on domoic acid to link hippocampal damage to seizures and memory impairment.

But the thylacines were gone, and nobody had studied their behavior while they were around. If I was to make sense of their brains and what it was like to be a thylacine, then I first had to understand the environment in which they had lived. I had to experience it myself.

I had been warned that the bush was thick, but it's hard to appreciate what that means until you're in it. Australians use "bush" to describe pretty much all types of wilderness. On the mainland, the bush is mostly desert. But in southwest Tasmania, as I was quickly learning, the bush is rain forest and jungle.

I ducked beneath a stand of pandani palms, and the serrated edges tugged at my clothes, as if the forest were trying to prevent interlopers from gaining access to its secrets. I pushed aside the leaves, and water from the previous night's rain trickled down my neck. The track, although visible, had not seen much traffic. The bush had a tendency to quickly reclaim the land.

The palms soon gave way to a moor of buttongrass. Sedges growing in mounds two to three feet high formed a lumpy landscape. And although not as high as the pandani, the buttongrass was no less effective in impeding my progress. The spiky grass, straight out of the Cretaceous era, thrives in soggy soil like that of the Tasmanian bush I was walking through. Not only did each step meet with resistance, but each ended with my foot sinking into an inch of water. I wouldn't find out until much later, but leeches were working their way into my boots.

Just a week before, I had been scanning another thylacine brain with Ken Ashwell at the University of New South Wales in Sydney.

On scan day, Ken had picked me up at my hotel and we had driven to the museum. Sandy Ingleby, the curator of the mammal division, was waiting outside. She held a box of champagne.

Ken joked, "Are we celebrating early?"

Sandy laughed and said, "It was the only box I could find to hold the specimen."

At the MRI room, Sandy carefully took the brain out of the champagne box. The smell of alcohol was strong.

"What's it in?" I asked.

Nobody knew.

Sandy said, "Ken, you can do the honors."

Ken put on some gloves and carefully lifted the brain out of its jar. He laid the brain on a scale. It weighed 30 grams.

"That's twice as much as the Smithsonian specimen," I told

them. "That means it hasn't shrunk as much. We should get better signals."

The brain was by no means perfect. As I recorded the weight, I noticed a big gash on the top of the brain. That could be a problem. There was no way to know how deep it went before we scanned it. Also, the olfactory bulbs had been chopped off.

Since the clock was ticking, we had no time to immerse the brain in inert fluid. We just sealed it in a plastic bag and loaded it into the scanner. The 9.4 Tesla magnet was considerably smaller than the 3 Tesla one Peter and I had used at home. Such a high field was difficult to generate uniformly in a human sized magnet, so this one had a bore of about 30 centimeters in diameter, which is about half the size of the bore in an MRI machine designed for humans. It was big enough for rats and monkey brains. And thylacines.

The initial scans looked promising. The images had good contrast, which meant that the gray and white matter hadn't deteriorated into a uniform mush. And stronger magnetic fields meant that we could scan at finer resolution. We set it to 200 microns—one-fifth of a millimeter. At this resolution we would be able to see things just barely visible to the naked eye. It would be like putting the brain under a microscope, except we wouldn't have to slice it up, and we wouldn't be limited to two-dimensional images. The scan would take three hours. The diffusion images would take another three.

Satisfied that the brain was safe in the magnet, Sandy returned to the museum. We were, however, under strict orders not to leave it unattended.

The thylacine's foraging habits were largely unknown. Like all the marsupials, they were nocturnal, so few people had ever observed them hunting. As the apex predator in Tasmania, presumably they had hunted smaller animals, such as walla-bies, possums, and pademelons. Whether they were actually sheep-killers remains unclear.

Certainly it is possible that thylacine took sheep. They were similar in physical size to coyotes, who are well known, and despised, among sheep-farmers. In the United States in 2014, 61,712 sheep and lambs were lost to predator activity, and coyotes accounted for 54 percent of these deaths. But the thylacine's leg structure was different from that of coyotes, and therefore the thylacine must have had a different hunting strategy. Analysis of the thylacines' elbow joints has shown that they did not have the biomechanical structure for pursuit. They couldn't run fast. Instead, they would have evolved to ambush their prey.

A comparison of thylacine teeth and the teeth of other animals leads to similar conclusions. From the shapes of teeth, especially canine teeth, one can deduce how a carnivore kills its prey. Dogs have canine teeth that are not as wide side-to-side as they are front-to-back. Such flattened cross-sections make their teeth well-suited for tearing and slashing. Cats have teeth with rounded cross-sections, making them better suited to killing by puncturing and crushing. In cross-section, the thylacines had ovoid canines, intermediate between those of cats and canids, but similar to those of foxes and hyenas.

The one point of agreement between lore and modern reconstructions is that the thylacine was not fast. The early settlers and bushmen thought they were dim-witted, which would seem to be at odds with their reputation for killing sheep. But of course slow-moving does not mean stupid. I hoped the thylacine's brain would shed some light on the controversy.

At the least, our patience was rewarded with some beautiful images of the structure of the thylacine's brain. Because the

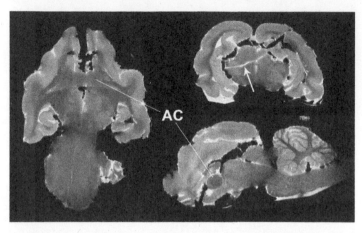

Three views of the Sydney thylacine. Scanning at 9.4 Tesla resolved details as small as 200 microns, but a gash had sliced through part of the brain (*arrow*). The anterior commissure (AC) is the major bundle of fibers connecting the left and right hemispheres. (*Gregory Berns*)

Sydney specimen was in better shape than the Smithsonian's, and because we had scanned at 9.4 Tesla, the resolution was astonishing. I could make out white matter tracts that we hadn't seen in the Smithsonian specimen. But the gash I had noticed when we put the specimen in the scanner went much deeper than I had expected. Whatever knife had caused it had cut a diagonal swath from the right cortex to the left thalamus. That would make it difficult to trace the white matter pathways, because neither hemisphere had been left unscathed.

With data from two of the four known thylacine brains, we were in a much better position to interpret the findings than we had been with only the Smithsonian specimen. With two brains, we could determine what was common to both, potentially filling in gaps due to damage in one or the other. But to make sense of the pathways, we still needed something to compare to the thylacine, preferably another carnivorous marsupial.

Anticipating this, I had started that effort even before going to Australia.

There aren't many carnivorous marsupials, and most that do exist are classified into the order Dasyuromorphia, which means "bushy tail." Besides the thylacine, the other dasyuromorphs include the quolls, the dunnarts, the numbats, and the Tasmanian devil. According to reconstructions of the family tree using mitochondrial DNA, numbats, also known as marsupial anteaters, are the closest living relative to the thylacine. But numbats are small insectivores. Their ecological niche is quite unlike the thylacines', which suggests their brains might be, too. Dunnarts are also insectivores, and are only the size of a mouse. Quolls (formerly called native cats), which eat lizards, birds, and other small mammals, make a better ecological match, but even the largest, the tiger quoll, only grows to about seven pounds. I had to eliminate all of them. That left the Tasmanian devil.

Devils do not have a great reputation. They mostly feed on carrion, which makes them as repulsive to many people as vultures. They eat everything, including bones, powered by the strongest bite strength per body weight of any land mammal in the world. They are loud, making a wide range of screeches, which can be heard through the night. And because they are mostly solitary creatures, they tend to fight with each other. This pugnaciousness doesn't win them many fans. It also threatens to wipe them out.

Since the late 1990s, devils have been dying in record numbers from a kind of cancer that grows on the face. These tumors get so big that eventually the animal can't eat and dies from starvation. It is called devil facial tumor disease (DFTD), and it spreads when the devils fight. Apart from threatening the entire species, the disease is of great interest because it is one of the few known instances of transmissible cancer. Unlike the human papillomavirus, which is associated with cervical cancer, DFTD isn't caused by a virus. Instead, it seems to be caused by the direct transmission of cancerous cells—in this case, cells derived from a Schwann cell, which is a type of cell surrounding neurons. DFTD is one of only four known instances of direct transmission of cancer. Dogs, clams, and Syrian hamsters are the hosts of the others. In dogs, the cancer is transmitted sexually, and it is called canine transmissible venereal tumor (CTVT), but it is not fatal like the devil cancer.

By 2008, the situation had gotten so bad that many conservationists believed Tasmanian devils would soon be extinct. Some regions of Tasmania had seen a 90 percent decline in the wild population. Recognizing the seriousness of the situation, the devil was declared endangered that year by the International Union for the Conservation of Nature and Natural Resources (IUCN).

At that point a consortium of scientists, conservationists, and government officials agreed to devote the resources necessary to save this iconic animal. The Taronga Conservation Society, which manages the zoo in Sydney, in collaboration with the Tasmanian Department of Primary Industries, Parks, Water and Environment and the Conservation Breeding Specialist Group (part of IUCN), came up with a plan to create an insurance population of devils. It was an enormous undertaking with complex logistics. The idea was to create a stable breeding population in captivity that could be released if wild populations disappeared. Similar programs had been put into place for only a few animals, notably the California condor and the red wolf in

North Carolina, and even then with mixed success.

In the event that wild devils went extinct, the working group estimated that a captive population of five hundred breeding devils would be needed to repopulate. These devils were housed in a network of zoos and wildlife parks throughout Australia. There were also a few free-range enclosures in Tasmania, which were less intensively managed environments that were designed to help the devils retain their natural behaviors.

Carolyn Hogg, a geneticist and conservation biologist at the University of Sydney, managed much of the breeding program. After Peter and I had scanned the Smithsonian thylacine, I had written to Carolyn about the possibility of obtaining the brain of a Tasmanian devil for comparison.

This was a sensitive issue. Devils are endangered, and there was a certain amount of national pride at stake. Since the disappearance of the thylacine, the devil had become the new icon of Tasmania. I had sensed some reluctance of the Australian authorities to give me any devil material. Carolyn, though, saw the value in mapping the devil brain, just as we were trying to do with the thylacine. She cut through the bureaucracy and found the brain of a recently euthanized devil to send me. The number of permits we had to obtain was daunting. The US Fish and Wildlife Service, which cleared the import of animal material, had never heard of anyone bringing a devil brain into the United States. The only way to be sure I got it intact was to arrange for a courier to meet the package at LAX and shepherd it through customs onto a domestic flight to Atlanta.

The devil brain showed up in October 2015, and we immediately prepared it for scanning. Unlike the thylacine brains, or even the dolphin brains, the devil specimen was relatively fresh. Even though it was small, we had no trouble getting good signals from it. The structural images were beautiful. The diffusion images looked just as good. Like the thylacine, the devil did not have a corpus callosum. The anterior commissure, though, appeared as a thick band of fibers connecting the hemispheres. I had high hopes that the devil brain would provide the comparison we needed to interpret the thylacine data.

Even though they were both carnivorous marsupials, the thylacine did not feed the way a devil does. That, according to some ecologists, should be evident from their brains. One theory suggests that animals with more complex feeding strategies need bigger and more complex brains than animals with simpler feeding strategies. Certainly this would be the case for thylacines, at least if they had been the sheep-killers that the bushmen had said they were.

But where to look? There wouldn't be a "hunting" region of the brain. Instead, as a measure of cognitive sophistication, we might be able to determine how much of the brain was "frontal"—the part of the brain in front of the motor regions. If the marsupial brain and the brains of placental mammals were organized in similar ways, then this region would be associated with cognitive tasks like planning, impulse control (as we saw with the dogs), and social processing.

Much more was known about the organization of placental mammals' brains than was known about the organization of marsupial brains. The discrepancy was surprising, because marsupials, especially the Australian ones, were thought to be similar to the earliest crown mammals of 150 million years ago. The organization of marsupial brains should provide a window into how mammalian brains evolved.

At the most basic level, one would like to know how sensory and motor information is organized. In placental mammals, a sulcus divides the primary sensory and motor regions, with a modular structure and multiple maps of the body within the brain. Marsupials, because they were more ancient, might have simpler brains. The earliest electrical recordings of marsupial brains suggested that they didn't have separate representations of motor and sensory information. Recent experiments, though, had painted a more nuanced picture. Leah Krubitzer, a neuroscientist at the University of California at Davis, has suggested that "primitive" marsupials, such as the Virginia opossum and the red kangaroo, had overlapping motor and sensory representations, while more "advanced" marsupials, like the wombat, the pademelon, and the devil, had separate representations.

It is, of course, tricky to define "primitive" and "advanced." It's better to look at the animal's sociality and feeding habits. In general, the more flexible the foraging strategy, the more complex the brain would have to be.

B etween the Smithsonian, the Australian museum, and the Save the Devil Project, we had accumulated data on four specimens: two thylacines and two devils. Only the brain of the recently euthanized devil was in good condition. All the others had suffered some degradation or trauma over the preceding century. It was very much a forensic reconstruction from disparate pieces. But the fact that we had managed to get two of each species helped a great deal. We could finally say what was common in each species and what was different.

Ken wanted to focus on the connections between the thalamus and the cortex. The thalamus contained dozens of different neuron clusters (called nuclei), but only the largest ones could be made out on the MRI images. He could make a good guess at their locations, but because they were buried in a larger structure, the white matter tracing would be subject to a lot of uncertainty, especially where they entered the thalamus.

For confirmation, I focused on a collection of structures known as the basal ganglia. A lot was known about the basal ganglia. In fact, we had been studying a major chunk of it in the Dog Project—the caudate nucleus. The basal ganglia were large and easily identifiable, so it was easy to trace subparts of it even in the strange marsupial brains. This was critical. Most of the differences between brains seemed to be in the size and layout of the cortex. The so-called subcortical structures, such as the basal ganglia and the thalamus, seemed to be relatively invariant across species.

The idea was to see how the subcortical structures were connected to the cortex, which would give us a high-level map of both the devil brain and the thylacine brain. The nice thing about the basal ganglia is that, at least in placentals, parts of it are known to be associated with motor functions, while others are more sensory or cognitive. In placental mammals, like dogs and humans, the front of this arch is connected to parts of the frontal cortex associated with cognitive functions such as planning and impulse control. In the other direction, the tail of the caudate is matched to parts of the cortex that process sensory input. Moving laterally from the caudate, another piece of the basal ganglia, called the putamen, is known to be connected to motor systems.

For the thalamus, Ken located the nuclei for transmitting visual information and—as we had done with the dolphin brains—auditory information. He also located the nuclei for tactile inputs and motor outputs. When we reconstructed all of the pathways to the cortex and integrated them with the basal ganglia pathways, a coherent picture emerged.

In broad patterns, the devil maps and the thylacine maps looked similar. The most rearward part of the cortex was predominately sensory, and this corresponded to the visual and auditory regions. Moving forward, we found a large chunk of cortex devoted to motor functions. This arrangement—motor in front of sensory—seems common across mammalian brains and is similar to what is seen in rats, dogs, and humans. In front of the motor region, we saw a likely cognitive area, and then the most forward, probably an emotion/motivation region.

The thylacine, though, seemed to have relatively more "stuff" in front of the motor region. This was the smoking gun I had been looking for. In all the other mammals, species with more complex environments have an expanded prefrontal cortex. Such complexity is often seen in predators, especially those that have complex social structures. The sea lion brains Peter and I had studied had more prefrontal cortex than those of harbor seals. Although they both eat fish, sea lions have a more elaborate foraging strategy, often venturing deeper into the ocean and farther from shore.

By virtue of its larger prefrontal cortex, we could conclude that the thylacine had a more complex mental life than its closest living relative, the Tassie devil. Did the thylacine have enough of a prefrontal cortex to achieve sentience? Considering that we know very little about how much of a brain is necessary for self-awareness, that is a difficult question. But based on the size of the frontal cortex and its connections to other parts of the brain, I was convinced the thylacine was an intelligent, emotional creature.

The marsupials have never gotten the credit they deserve. Because they are considered more "ancient" than placental mammals, the marsupials are often assumed to be less evolved, and therefore less intelligent. But this is silly. The marsupials have had as long an evolutionary history as any other mammal, including us, and until recently, they ruled the Australian continent with much greater diversity than we see today. There even used to be giant marsupials. The diprotodont was a hippopotamus-sized wombat that lived until about 40,000 years ago. Like today's wombats and the koala, they were herbivores. The king of predators, though, would have been the "marsupial lion" (*Thylacoleo carnifex*), a kind of jacked-up thylacine. About the size of a modern-day lion or tiger, the marsupial lion was well-adapted to hunting. Its massive jaws with large canines would have delivered fatal puncturing bites. The marsupial lion is estimated to have had the largest bite strength per body weight of any mammal that has ever lived.

Until about 12,000 years ago, Tasmania was connected to the mainland. When the glaciers retreated at the end of the last ice age, the sea level rose, and Tasmania became an island unto itself. Whatever plants, animals, and people already inhabited the island were then left to coexist on their own. By about 4,000 years ago, any thylacines left on the mainland had disappeared, probably as a result of ongoing climate change and competition from Aboriginals and their dogs. This left a population of perhaps a few thousand thylacines in Tasmania.

By the time the British arrived, there were probably 1,000 or so Aboriginals in Tasmania. Much as the interlopers declared war on the thylacine, they did on the Aboriginals as well. Although the thylacines couldn't fight back, the Aboriginals did. The Black War, which began in the mid-1820s, is largely unknown outside of Australia. The outcome was the eventual

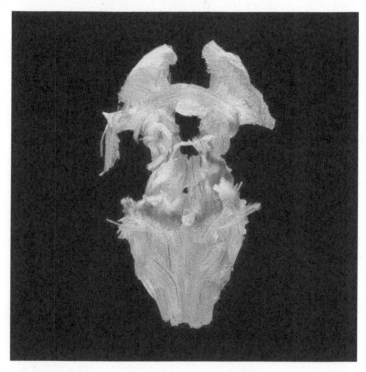

Visualization of fibers in the brain of the Tasmanian devil shows upsweep of cortical fibers. (*Gregory Berns*)

eradication of the Aboriginal population. Because the same colonists also destroyed the thylacine population, the events leading to the Black War also shed light on the disappearance of the Tassie tigers.

The majority of settlers didn't even want to be in Tasmania. Convicts and soldiers alike had little interest in the natives, and the scarcity of female colonists fostered an environment of abduction and rape of Aboriginal girls and women. The colonial attitudes toward indigenous people reflected the general mindset of surviving in a difficult frontier and the prevailing assumption among the British that they had a God-given right to be there.

Had rape not been so common, a peaceful coexistence might have been possible. Unsurprisingly, Aboriginal tribes carried out acts of revenge. By the time they realized that the colonists were increasing in numbers, war was inevitable.

Initially, many of the settlers, especially those in the main port of Hobart, sought a humanitarian resolution—peaceful coexistence, or perhaps relocation of the Aboriginals to an island. But those on the frontier thought differently. In May 1828, the *Colonial Advocate* argued that anything short of extermination of the Aboriginals would be the height of absurdity. By 1830, roving parties were searching the countryside. Some of these parties were made up of military and field police, but others consisted of convicts who were hunting for blacks on the promise of a reduced sentence.

Nicholas Clements, a historian at the University of Tasmania, astutely identified the change in attitude toward the Aboriginals, writing, "Disinterest in the natives shifted to distrust in the initial years of the war, and thereafter, to hatred and fear."

The Aboriginals fought back. As they lacked the rifles of the Colonials, they resorted to two tactics in particular: arson and stock-killing. The killing of livestock was important because the Aboriginals never took the carcasses. The colonists kept a close eye on their horses, so the Aboriginals directed their attacks on sheep and cattle, killing thousands during the war.

And this is where the story sheds light on the fate of the thylacine.

The first reports of sheep-killing due to thylacines date to 1824—about the same time the conflict with the Aboriginals was heating up. Notably, thylacines were reported to kill sheep and drink their blood but leave the carcasses. Never mind that blood had little nutritional value to an apex carnivore, or that major predators do not leave their kills behind for other animals to eat. It became accepted wisdom that the thylacines were the primary cause of sheep losses.

The Van Diemen's Land Company, which held rights to vast tracts of land, including the entire northwestern portion of the island, bore the brunt of these livestock losses. Whether the cause was thylacines, Aboriginals, or wild dogs, the mounting losses of sheep caused the company to introduce a bounty system in 1830. Five shillings were paid for every male thylacine and seven for females, which would have been enough to rent a single room in London's Soho district for a week. It would have been worth even more in Tasmania. Devils and dogs commanded only half the price of a thylacine.

Eric Guiler, who before his death in 2008 knew more about thylacines than anyone else, could never ascertain how much of the sheep-killing had been actually caused by thylacines. Some of the bushmen he interviewed during his career swore they had seen a thylacine killing several sheep in one night. Others, however, thought that most thylacines had been uninterested in sheep and would walk through a flock without disturbing them. Even during the bounty era, the people working for Van Diemen's Land Company knew that wild dogs were probably responsible for a good portion of the losses. This would be consistent with the hunting styles of canids, like the modern-day coyote. Nevertheless, the thylacines continued to be blamed for the losses.

The dog theory was also supported by Clements, who

thought that dogs, which were imported by the British themselves, were the primary killers of sheep. Clements pointed out
that some of the dogs were also adopted by the Aboriginals
for hunting and companionship. In an email, Clements wrote,
"During the war, it was dangerous to keep fires because this was
the primary way colonists found and killed Aborigines, so they
would snuggle with dogs for warmth." The dogs followed the
tribes and are thought to have killed great numbers of sheep.

By 1832, the Black War was over—at least in eastern por-
tions of Tasmania. The few surviving tribes there surrendered
and eventually relocated to Flinders Island, their numbers diminished to less than one hundred. But the conflict continued
until 1842 in Tasmania's northwest. Unlike the war around
Hobart, the war in the northwest was fought by farmers and
shepherds. The motivations, however, were the same: when the
overwhelmingly male colonists abducted the Aboriginal women,
the tribes retaliated.

Not coincidentally, the myth that thylacines herded sheep
off cliffs also began to circulate around this time. The northwest was remote, even by Tasmanian standards. The farthest
point is still called Cape Grim. The colonial shepherds and rep-
resentatives of Van Dieman's Land Company operated under an
extreme profit motive, without much legal restraint. It is easy
to see how thinking the Aboriginals and the thylacines were
responsible for the sheep losses ultimately led to the destruction
of both.

The intertwining of Tasmania's colonial past with the fate of
Aboriginals and thylacines made a lot of sense to me. But
it didn't say anything about thylacine behavior, other than that

they weren't the sheep-killers most people thought they were. And there was nobody alive who had seen an actual thylacine. Except, perhaps, Col Bailey, the last thylacine hunter and true believer.

Col wrote a famous account of an encounter with a thylacine in 1995 in Tasmania's remote southwest, in the Weld Valley near the Snake River. Even in 2016, this was one of the most difficult-to-reach regions of Tasmania. If the thylacine had survived, it would probably have been in this region, which was still relatively untouched by humans.

I met Col at his home about an hour outside of Hobart.

Col and his wife, Lexia, used to live in the wild reaches of the southwest, from which Col pursued his passion of thylacine tracking. But now, in their seventies, Col and Lexia had downsized to a small house in town with more conveniences than the frontier provided. Col had converted the sitting room into an office, which was chock full of thylacine memorabilia.

He opened up one of the countless notebooks and showed me a picture of a paw. "This is the foot of a thylacine. I took this picture in the home of the guy who killed it in 1990."

I didn't know what to think. It looked like a dog's paw. But I had no knowledge of how to differentiate the paws of canids and carnivorous marsupials.

My skepticism must have been obvious, because Col pulled out a similar picture of a foot taken by Stephen Sleightholme. It was a specimen from the museum at Oxford University.

"See," Col said. "Identical."

"Where was the thylacine killed?"

"Near Adamsfield."

In the late nineteenth and early twentieth centuries,

Adamsfield had been a mining town. Miners had flocked there to search for osmiridium—an alloy of osmium and iridium—which was much rarer and more valuable than gold. Adamsfield was abandoned after World War II, and the bush had long reclaimed the area.

Adamsfield was also where Elias Churchill had trapped Benjamin in 1933.

"If you want to see Tiger country," Col said, "go to Adamsfield."

This was why I had come. I didn't expect to discover any thylacines, but I wanted to see where they had lived. I needed some frame of reference for understanding their habitat and how their brains had evolved in response to a particular environmental niche.

"What do you think?" Col asked.

"About what?"

"Are they still out there?"

"I don't know," I said. "The odds are against it."

Col shrugged. He had heard it all before. Skeptical scientists.

I said, "You'd think that with all the improvements in technology, someone would have captured one on camera."

"Thylacines are shy," Col replied. "They have an incredible sense of smell and they would stay away from anything that smelled like a human."

I was still skeptical. Driving to Col's house, I had seen plenty of wallabies splatted on the road. There was no shortage of thylacine prey, so all I could conclude was that none were around. At least in this area.

"Suppose," I said, "you did find proof that they still existed. What would you do?"

"Ah, you've hit the nail on the head. I suppose I would keep it to myself. Maybe I'd get a piece of hair to send to Mike Archer to clone."

I sensed that Col's attitude was common. Distrust of the government ran deep in this part of the country, and most people thought that the authorities would just mess everything up.

At least I learned that Adamsfield was where I needed to go. Getting there was another matter. Since there weren't any roads into that part of the wilderness, Col couldn't give me exact directions. But following a hunch, the next day I checked in with the rangers at Mount Field National Park.

Mount Field was a gateway to the Tasmanian Wilderness World Heritage Area, which was one of the last great wildernesses on Earth, ranking with the Serengeti in Africa, or the Redwood National Park in California. The plants and animals in the heritage area were utterly unique and worthy of protection. Most thylacine enthusiasts believed that if any survived, they would be in the World Heritage Area.

I chatted up an athletic-looking ranger at the visitor center.

"So how do I get to Churchill's hut?"

The ranger didn't seem at all surprised by the question, although she looked at me for a beat. Concluding that I was probably fit enough to make the journey, she printed out the directions.

She said, "I was there a couple of months ago. I put some tape on the track, but I don't know if it's still there."

I thanked her for the directions and signed the log book.

Another ranger perked up at this point. "We don't check the log book unless someone reports you missing."

"But I'm here by myself. Nobody would know if I go missing."

"Is there anybody you can tell where you're going?"

"I'm telling you."

They seemed a bit uncomfortable with this but agreed that I could check in with them on my return. If I didn't make it back by the end of the day, they would come looking for me.

From the logging track to the mysterious road "No. 5," the ranger's directions were spot-on. Judging from the condition of the trail, she might have been the last person to walk it. It had rained the night before and every step was met with a sinking squish into the muck. The bush was dense and pulled at my clothing. If there were thylacines around, they could have been watching me from just off the track, and I wouldn't have noticed.

The buttongrass grew in mounds to waist high. A thylacine's stripes would have blended in perfectly with the spiky leaves and the straight shadows they cast. The soft ground would have muffled any footsteps, and I imagined the creatures moving silently through the bush. They would have moved deliberately, scenting their prey with their large olfactory bulbs. Indeed, it would have been hard to hunt by sight in the bush. They would have positioned themselves strategically downwind of their prey and pounced on them when they came near.

As I pressed on toward Churchill's hut, it was easy to imagine what it was like to be a thylacine.

Lonesome.

The thylacine's prefrontal cortex was sufficiently large to support complex cognitive operations, perhaps even large enough for self-awareness. But just because a thylacine was solitary didn't mean it would be lonely. A male like Benjamin wouldn't have pined over his lonesomeness. Females would have had a maternal bond to their young, but even that would have been short-lived. As soon as they were capable, the pups would have left the den to fend for themselves. It is hard to say whether they would have recognized each other if they met later.

Compared to the other carnivores we had studied—dolphins, sea lions, dogs—the thylacine would have been all business. Lacking a strong evolutionary drive for sociality, they would not have made good pets. Their stoic disposition gave them a reputation for being slow-moving and dim-witted. In actuality, the thylacine was rather asocial and shy. But not stupid. There are no stupid carnivores. Like all of the other predators we had studied, the thylacine had a brain and a mind capable of outsmarting the animals it hunted.

I soon reached a point where the track was cut by a rather wide and fast-flowing creek, swollen by the recent rain. I could barely make out a primitive structure on the other side—Churchill's hut. My boots were already waterlogged and leech-filled, but given my solitary situation, I decided that it was unnecessary to attempt the creek crossing.

A thylacine wouldn't have been deterred.

But I was still a social creature, and my family called for my return.

Chapter 11

Dog Lab

Ever since we had started the Dog Project in 2011, three principles had guided our research. All of them went beyond what was required of researchers working with animals.

First, we would never do anything to harm the dogs. This meant using only positive reinforcement to train them. It also meant we wouldn't cause the dogs any pain. That may seem obvious, but I received several inquiries from veterinarians about the possibility of using fMRI to study pain in dogs. Pain management is an important issue in animal care, but I could not see any ethical way to inflict pain on the MRI-dogs, even if the knowledge gained might help other animals.

Second, we would not restrain the dogs. This meant no physical restraint and no sedation. Part of the rationale was scientific. After all, you can't learn much about the brain of an unconscious animal. And forcing animals to participate by strapping them down would only cause anxiety. But the main reason was ethical. We would never force people to participate in research, so why would it be okay to force animals?

This question led to the third, most iconoclastic principle. We gave the dogs the right of self-determination. We provided steps for them to walk into and out of the scanner instead of placing them on the patient table. Above all, we treated them as sentient beings with their own likes and dislikes. Because we did all these things, the dogs had the same fundamental privilege as humans participating in research: the right to refuse.

By doing these things, especially by granting the dogs self-determination, I felt like I was committing one of the great cultural and scientific sins. It was heresy to allow an animal the choice of submitting or not submitting to a human's will. It rejected practices typical of the entire industry of laboratory animal research, not to mention industrialized farming.

But, I am ashamed to admit, I have not always behaved this way.

In 1990, when I was in medical school, the aspiring doctor had to navigate two great rites of passage: gross anatomy and dog lab.

For most of my classmates, gross anatomy was the capstone experience of the preclinical curriculum. While the first year of medical school was devoted to learning about normal human bodily function, the second year introduced us to the effects of disease. Gross anatomy straddled the divide between health and pathology. For many, it was the first time they had seen a dead body, and for the most sensitive, the lab was approached with great trepidation.

The faculty, though, had been doing this for decades and knew that the key to learning anatomy was to distance oneself from the obvious fact that the specimen on the table had, not too long before, been a human being. It was easy to forget this because the most human part—the head—was saved for last. We spent most of a semester dissecting everything below the neck, and only at the end were we allowed to unwrap the head and learn the secrets of everything inside of it.

Some of my classmates struggled with the dissection of a fellow human being. I did not. I was seduced by the beauty of the human body. Even in death, the perfection of this machine was unrivaled. And the people who became these vessels of our education had done so of their own choice. We could never know what motivated someone to donate his or her body for such purposes. A desire to turn one's death into something that would help others? A final, dignified act. The end result turned out to be a life-altering experience for many. I left with a new appreciation for the complexity of the human body, as well as the confidence to cut into one, which would happen soon enough during surgical training.

But if gross anatomy could take the result of death and elevate it to something worthwhile and even profound, dog lab did the opposite.

Dog lab was supposed to demonstrate how various drugs affected cardiovascular function. Everything in the lab centered around the Frank-Starling law, which stated that the more blood that returned to the heart, the harder it would pump. For the first time, we would be able to administer real drugs to living creatures and see how their heart rates and blood pressure changed. We all accepted the implicit justification that doctors had to learn on animals before people.

Even though I had spent an entire semester with my cadaver in gross anatomy, I don't remember any details about the body. Some groups had named their cadavers, but my group did not. I can't even remember if it was male or female. But dog lab, which lasted all of one morning, burned into my memory, periodically surfacing to remind me of what I did.

There were 120 students in my class, and we had been divided into teams of 4. We entered the physiology lab wearing our crisp white medical-student coats. On each of the thirty stainless-steel tables, a dog was laid out on its back. To keep the dogs in position, their paws had been tied down, with each paw secured to a corner of the table. The dogs were already anesthetized. My group's dog was a female with a short wiry coat in patches of white, black, and tan.

The professor instructed us to periodically monitor the depth of anesthesia by firmly pinching the webbing between the toes. If the dog didn't try to pull the paw away, the anesthesia was deemed adequate. My hands were cold from nerves. When I stroked the dog's paw, I recoiled at how warm she felt.

The lab protocol required us to inject a series of drugs. Epinephrine (adrenaline) increased the dogs' heart rates and blood pressure, while acetylcholine had the opposite effect. After we measured the effects of several drugs, the professor's surgical assistants opened the dogs' chests so that we could see the heart beating and the lungs inflating. When we were done, the final step of the protocol called for an injection of potassium chloride directly into the heart, which would cause it to stop beating.

For decades potassium had been used as the final drug in the cocktail for executions. Although potassium would stop the heart, it wasn't instantaneous. It did it by slowing the heart down. Even worse, it didn't always work right away. We would have to let ten minutes pass without a heartbeat to be sure it had stopped.

When we reached that point, the professor supervising the

dog lab came over to our table and said, without a trace of emotion, "It's quicker to cut the pulmonary artery."

None of us wanted to stand a grim vigil waiting for the potassium to work, but none of my labmates wanted to kill the poor dog, either. The professor was not going to do it for us. He crossed his arms while we decided what to do. This was a rite of passage, and we had to face death.

So, I took the autopsy knife from the professor's hand, lifted up the heart, and severed the vessels behind it. Warm blood flooded the chest cavity. Without any blood to return, the heart deflated and stopped instantly.

The professor nodded and walked to the next table.

I have never told this story before. It is one of the deep regrets of my life, and I wish I had been strong enough to boycott the lab. The thought had crossed my mind, but I justified participating with the standard reasons: the dogs came from the pound and would be killed anyway, and doctors needed to see how real living systems worked. In retrospect, I realized neither was true. The dogs wouldn't necessarily have been killed, and the laboratory exercise only confirmed what we had been taught in class. Seeing how the drugs worked on a live animal didn't impart any additional knowledge.

The lab didn't make me a better doctor, and it diminished me as a human being. The practical knowledge that was supposedly gained at the expense of the dog I later learned more directly and much more truly by treating humans in clinical settings. I think now that, by trying to figure out what dogs think and feel, I was trying to make amends. If I could prove that their subjective experiences were similar to ours, then pointless academic exercises like dog lab would be unjustifiable. The last US medical school to use live animals in training finally stopped in 2016. The shift away from using animals in medical schools had little to do with the Dog Project. Instead, growing pressure from the animal-welfare movement, organizations like People for the Ethical Treatment of Animals (PETA) and antivivisection societies, along with changing attitudes toward animals caused medical schools to abandon live animals for training purposes. Computer simulations had also achieved a level of reality that made it hard to justify killing animals.

Jeremy Bentham, the English philosopher, is most frequently credited with launching the modern animal-welfare movement. Bentham famously wrote in 1780, "The day may come when the rest of the animal creation may acquire those rights which never could have been withholden from them but by the hand of tyranny. . . . The question is not, Can they *reason*? nor, Can they *talk*? but, Can they *suffer*?" Bentham was a utilitarian. As such, he was concerned with objectifying the outcomes of actions. Actions that increased well-being and happiness, or that decreased pain and suffering, were desirable. This philosophy led to the principle of the greatest good for the greatest number.

Although Bentham is credited with recognizing that animals could suffer, the blunt calculus of utilitarianism still subjugates animals' interests to humans', because a human life is always deemed more valuable than an animal's. This is why most people implicitly accept that it is okay to kill animals as a source of nourishment, or to make clothing, or to advance medical science. Under the principle of greatest good, such actions are not only justifiable, but desirable. Although the Cruelty to Animals Act was passed in the United Kingdom in 1849, it wasn't until 1966 that a federal law regulating the suffering of animals was passed in the United States—the Animal Welfare Act, or AWA.

Originally, the AWA was meant to prevent pets from being stolen for sale to research laboratories and to regulate how dogs, cats, and other animals should be handled in research. The act has been amended several times since to prohibit things like animal fighting and to expand the list of practices considered cruel to animals. Still, the AWA was directed primarily at animal research and did not apply to all animals. It defines animals as "any live or dead dog, cat, monkey (nonhuman primate mammal), guinea pig, hamster, rabbit, or such other warm-blooded animal, as the Secretary may determine was being used, or was intended for use, for research, testing, experimentation, or exhibition purposes, or as a pet." The act excludes birds, rats, mice, horses not used for research purposes, and other farm animals used for food or fiber.

The act specifies that universities must establish a committee to oversee animal research. In theory, these Institutional Animal Care and Use Committees, or IACUCs, evaluate the appropriateness of animal research proposals. But the AWA didn't specify how the committees would determine what should be allowed. Instead, it took the utilitarian stance that animal research, by definition, benefited the greater good. The primary issues pertained to minimizing suffering.

A peculiar metric emerged. It is called the "three R's": replacement, reduction, and refinement. Replacement means that researchers should look for alternatives to using animals, either nonliving systems, like a computer simulation, or, failing that, the "least sentient" animals possible. Since nobody knows how to measure sentience, a natural order has emerged as a stand-in. Apes and dolphins are considered more sentient than dogs, and dogs more than rats, rats more than fish, and so on. The scale is highly arbitrary. Rats, for example, probably have as much personality and intelligence as dogs do. Reduction is pure utilitarianism. If you have to use animals, then try to use as few as possible. This also means you shouldn't needlessly duplicate experiments, which would unnecessarily cause more animal suffering and death. Refinement means that researchers should refine their methods to inflict as little pain and suffering as possible.

It is worth noting that a similar rubric had emerged for farm animals. In England, the Farm Animal Welfare Council, founded in 1979, began advocating for five freedoms for animals: freedom from thirst and hunger; freedom from discomfort; freedom from pain, injury, and disease; freedom to express normal behavior; and freedom from fear and distress. So even though farm animals would ultimately be killed for food, ethical farming recognized that animals had feelings and that human caretakers had the duty to make their lives as pleasant as possible.

On the surface, these principles sound reasonable, and they were revolutionary when they were written, but their impact has been limited. Even in the 1970s, as Peter Singer detailed in his book *Animal Liberation*, the three R's were routinely ignored. The basic problem was that they were guidelines, not requirements. Moreover, even if they were requirements, a utilitarian calculus of human benefit could always be used to rationalize the most horrid animal experiments. The five freedoms seemed to be taken seriously in Europe, but not so much in the United States.

Singer wrote that "pain is pain." This may seem obvious, but it is still not widely accepted by biomedical scientists. There is a long tradition stretching back to René Descartes, who wrote that animals were automatons that didn't think or feel pain, which continues to guide much scientific thinking. Even the modern principle of replacement implies that "lower animals" suffer less and that scientists should use the simplest animal possible to accomplish their goals. It was only in 2012 that a group of scientists wrote the Cambridge Declaration on Consciousness, which says, "Humans are not unique in possessing the neurological substrates that generate consciousness."

There is no ban on pain. Indeed, the US Department of Agriculture (USDA) only requires that research institutions categorize the level of pain. Category C is "no-pain," a category that includes euthanizing an animal that has been anesthetized. Category D is for pain that is relieved with drugs, and Category E for pain that is unrelieved by drugs.

In 2015, the USDA reported that 767,622 animals were used in research. Of these, 61,101 were dogs, and of the dogs, 40,071 fell into Category C, 20,668 in Category D, and 362 in Category E. What were all the other animals? Guinea pigs, hamsters, rabbits, monkeys, apes, cats, pigs, sheep, and others not specified in the USDA records. Guinea pigs and hamsters suffered the most, accounting for 80 percent of the animals in Category E. Because the USDA breaks the data down by state, it is easy to see where the animal experimentation occurred. For Category E, Missouri and Michigan led the pack with hamsters and guinea pigs, while New Jersey had the ignominy of leading in dogs. Considering all pain categories, New Jersey accounted for fully 13 percent of all animals in research. Why? Big pharma is heavily concentrated in New Jersey.

Nobody knows how many rats and mice are being used, because there is no federal agency that tracks them. Estimates range from 25 million to 100 million per year. This is striking, because the difference between a hamster and a rat is not so great. There is no ethical reason why one species should be covered under the AWA and the other not.

It was against this backdrop that we gave the dogs the option of not participating in the MRI research. Nothing in the AWA, the three R's, or the five freedoms said anything about self-determination. And yet, self-determination was the foundation of human medical research. It arose directly from the Nuremberg Trials. Self-determination was so important that the first principle of the Nuremberg Code was that of informed consent. People must know what they are consenting to, and they must be able to consent voluntarily.

We simply applied the same principle to dogs. Although the dogs couldn't understand the purpose of the MRI, we still granted them the right to refuse. And sometimes they did. Despite all the preparation and training, when confronted with the real MRI scanner, a few dogs were too anxious to go inside. We would work with these dogs to show them that the scanner was a big treat machine, but sometimes nothing would appease them. At that point, the owners usually realized that their dogs weren't cut out to be MRI-dogs.

What would happen if dogs and other animals were more broadly granted the principle of self-determination? Some might say it would mean the end of medical research, or that we would all become vegans. After all, what rat would consent to living in an electrified cage, only to have his head chopped off at the end of the experiment? What chicken would want to live in a square foot of space?

It is no surprise that most people accept the necessity of eating animals and using them for medical research. Some don't give it a second thought. Others, like me, struggle with their choices. Psychologists call it cognitive dissonance. Hal Herzog, a researcher of human-animal relationships, has said that it is impossible to be consistent with our treatment of animals.

In more pastoral times, animals lived on farms. They were fed and taken care of, living a good life right up to a swift and painless death. It might have been a reasonable tradeoff to be taken care of in exchange for feeding their caretakers. Now it's hard to know the extent to which farm animals ever lived a good life, but modern industrialized agriculture certainly does not meet any reasonable expectation of what animals want. The counterargument, of course, is the tired refrain that we don't know what it's like to be a farm animal.

But I think we do. Brain imaging, both in terms of structure and function, shows enough similarities that it is reasonable to extrapolate varieties of experience across a wide range of animals. With similar brain architectures for the experience of joy, pain, and even social bonds, we can assume that animals experience these things much like we do, albeit without the words for those subjective states.

The important question now is whether animals are aware of their suffering. If they aren't, one could make a case for eating them. But if they are, well, that changes the calculus. An animal who is aware of his or her own pain and suffering may well experience the existential fear associated with imminent death. And awareness of other animals' fear can only heighten such terror.

We do not yet have enough information about the neural basis of self-awareness to definitively answer the question, but there is enough circumstantial evidence to suggest that many animals are aware of their internal states, meaning they are sentient: rats can experience regret, dogs can value praise as much as food, and sea lions can do basic logic. Beyond sentience lies consciousness and self-awareness. Although it is the crudest measure, brain size may be a reasonable starting point for gauging consciousness. Bigger brains are more modular, which necessitates the coordination of information flow between regions. As far as we know, consciousness comes from this flow of information. Bigger brains, then, are likely to have higher levels of consciousness. Given that a cow's brain weighs about the same as a chimpanzee's, and most people would not eat a chimp, logically we shouldn't eat a cow either. But nobody knows how much brain is necessary for consciousness.

Neuroscience may not even be necessary to demonstrate self-awareness. A very clever experiment in 2016 showed that mice may be sufficiently self-aware to be fooled by a particular illusion, something previously known only in humans. The rubber-hand illusion was first demonstrated in 1998 by Matthew Botvinick and Jonathan Cohen, psychologists then at Carnegie Mellon University. The illusion is created by having a person sit with his or her left hand resting on a table. The hand is hidden from the subject's view by a screen, and a rubber hand is then placed in front of the subject in full view. While the person looks at the rubber hand, the researchers simultaneously brush both the real hand (hidden from view) and the rubber hand. Most subjects report that they feel a sensation as if they are being brushed in the location of the rubber hand, and 80 percent say that the rubber hand began to seem like it was their own. The rubber-hand illusion is thought to occur because our notion of self depends on the integration of multiple sensory inputs, especially vision and touch. For humans, these are two senses crucial for defining the boundary between the body and the environment. When visual and tactile signals conflict, as in the illusion, then the brain does the best it can to make sense of the information, including the extension of self to the rubber hand.

The experiment with mice in 2016 tested for the same illusion, but instead of a rubber hand, it used a rubber tail constructed from a piece of wire coated with synthetic hair. Approximating the human experiment, the researchers simultaneously brushed the mouse's real tail, which he couldn't see, and the fake tail. To train the mice, twenty-minute sessions were carried out daily. After a month, the researchers tested the mice by briefly grasping the fake tail. Remarkably, the mice turned their heads toward the fake tail as if it were their own, indicating that they experienced body ownership of their tails and that they could be tricked, like humans, into extending the ownership to a fake one. It was a small but important piece of evidence for self-awareness.

I am not so naïve to believe that neuroscience will result in a flurry of legal changes for animals. The legislative and judicial spheres are notoriously resistant to science, not so much because people don't believe in science—although many do not—but because laws generally reflect society's moral intuitions. Laws aren't made because science proves something. Laws are made because enough people believe something is right or wrong. That doesn't mean that science is irrelevant, however. Science can influence laws indirectly by changing people's moral sentiments.

When I wrote a *New York Times* op-ed, "Dogs Are People, Too," in 2013, I wanted to tell the story of the Dog Project. But I also speculated about the implications of finding evidence for emotional processes in the dog's brain that were similar to our own brains. I suggested that it would soon be time for a reconsideration of treating animals as property. I didn't expect that would happen a month later.

That November, Shannon Travis and Trisha Murray appeared in the Superior Court of New York of Judge Matthew Cooper. The couple had been married for less than a year when they decided to divorce. While Travis was away on business, Murray moved out of their New York apartment. She took some furniture. And she took Joey, their dachshund.

Travis believed that Joey belonged to her because she was the one who bought him from a pet store. Murray disagreed, arguing that Joey belonged to her because, among other reasons, it was in the dog's best interests. She gave the example that Joey slept next to her side of the bed. With little precedent to go on, Judge Cooper had to decide whether it made sense to hold a custody hearing for a dog.

It was one thing to adjudicate the division of property, but it was an entirely different matter to consider the interests of the dog. If the dog was property, as current law said, then it could have its own interests no more than a piece of furniture. That would be like saying that a chair cared about which room it was placed in.

In a remarkably readable opinion, Judge Cooper struggled with the concept of canine custody. Scouring case law, he found examples of courts ruling that pets were chattel and other instances in which they were something else. In 1979, a civic court in Queens had held that "a pet is not just a thing but occupies a special place somewhere in between a person and a piece of personal property." And in 2001, the Wisconsin Supreme Court had written, "Society has long since moved beyond the untenable Cartesian view that animals are unfeeling automatons and, hence, mere property."

Along these lines, Judge Cooper cited my op-ed, but it wasn't the neuroscience that directly impacted his decision. Although it might be scientifically possible, he didn't think it was practical to use MRI to "gauge a dog's happiness or its feelings about a person." To Cooper, the fact that my op-ed had appeared in the *New York Times* nevertheless indicated something: he thought the interest in our work reflected a growing sentiment that dogs (and other animals) were more than property.

Judge Cooper ruled in favor of a one-day custody hearing that would consider Joey's case according to the standard of what was best for all concerned, including Joey. Ironically, the couple didn't return to court. Instead, they reached an agreement in which Murray, the one who had taken Joey and whose side of the bed he slept near, had sole custody.

G etting people to rethink dogs is easy. Throughout the industrialized world, people spend money on their pets that would have been unheard of a few decades ago. Whether dogs and cats serve as status symbols or surrogate children, cultural norms have shifted. Recognizing that many dogs come from puppy mills, cities have begun to ban the sale of puppies in pet stores. Celebrities can be seen with rescue dogs instead of the breed-du-jour.

But what does neuroscience mean for other animals?

Our results have shown, no matter which animal's brain we examined, that if it has a cortex, the animal is very likely sentient, and that its subjective experience can be understood by degrees of similarity to ours. It doesn't matter if it is a bat or a dolphin, a sea lion or a marsupial.

Humans have always played favorites. That's why we have

racism, sexism, and speciesism. With time, however, all have diminished. Progress in human equality is spilling over into considerations for the animal world, too. Of course, it makes sense that such affordances would begin with the animals we live with, but the rise in popularity of non-animal food sources is just one piece of evidence that the concern for animals is spreading beyond the home.

In 2013, the National Institutes of Health (NIH) stopped funding research using chimpanzees and began retiring to sanctuaries the chimps that had been housed in the national primate facilities. In 2016, Ringling Bros. and Barnum & Bailey Circus, in response to increasing outcry, accelerated the phase-out of elephants from the circus. But without the elephants, the circus couldn't sustain itself, and it closed in May 2017. Sea World suffered devastating revelations in the 2013 movie *Blackfish*. Three years later, after ticket sales declined, Sea World announced it would stop breeding killer whales and curtail their use in shows.

The world's megafauna—its chimps, elephants, whales, and the like—have been the first to benefit from the increasing recognition of animal sentience. Although this is still speciesism, we should view it as a positive development. These large animals, along with some smaller ones, like the Tasmanian devil, that have had the fortune of being iconic, have become the ambassadors of the animal world. Any efforts to help them will help countless other species that share their ecosystems. Evolutionary history has shown that megafauna have disappeared when their ecosystems collapsed, from either natural or manmade causes. We can't track all the species on the planet, but we know where all the megafauna live. There are no places left to hide, so at least we know where we need to go to help.

Inevitably, the interests of these animals conflict with those of humans. The last of three northern white rhinos lived at great cost under twenty-four-hour protection in Kenya. At the same time, the people living around the rhinos were struggling just to survive. Some have argued that the resources being expended on these animals should be directed to helping humanity, such as finding a cure for AIDS. Have we gotten to the point that an animal's life is worth more than a human's? Should it matter if an animal is among the last of a species?

But these are the wrong ways to think about such problems. It is time to move past Bentham and utilitarianism. Utilitarian considerations ultimately fail because the animal and human domains are both bottomless wells of suffering. No matter how many resources we pour into either, there will still be more work to do. And it is always possible to tip the scale in favor of people because they are more valuable in the eyes of humans. The other extreme, based in the ethics of Immanuel Kant, is no better. Some people argue that all lives matter and that it is wrong to tally them up on a scorecard. A worthy goal, but not terribly practical for solving global problems.

Instead, I have begun to consider the role of animal advocates. Animals will never be able to argue for themselves. But neither can children or people who are mentally incapacitated. In those situations, where decisions must be made about a person, we often appoint a guardian.

Imagine what would happen if IACUCs were required to have an animal advocate, whose job it was to argue in favor of an animal's interests? IACUCs are already required to have at least one member from the community, whose job it is to represent the community's interests in "proper care and treatment of animals." An animal advocate would simply represent the animal itself.

Animal advocates could represent wildlife. Where habitats were being consumed for human activity, an advocate could argue on behalf of the animals who lived there. The same could be done for livestock. Crazy as this may seem, Judge Cooper's decision to allow a custody hearing for Joey was based on this idea. Although Joey didn't have an advocate, each party was set to argue on his behalf.

Of course, there isn't yet a court of law to hear arguments on behalf of animals. Because most jurisdictions still consider them things, animals don't have "standing." But this is slowly changing. Judge Cooper's decision is a case in point, and the Non-Human Rights Project, run by the lawyer Steven Wise, has been arguing on behalf of chimpanzees for courts to consider them as persons, which would confer a legal standing. And in 2016, Alaska, of all places, became the first state to pass a law requiring the consideration of animals' well-being in custody disputes, just like Joey.

Neuroscience isn't going to be able to tell us exactly what we should do, but it, and other technologies, will change what we know about animals' internal experiences. Brain imaging is one tool that will only get better. MRI technology continues to advance, pushing the resolution to finer and finer levels so that we will soon be able to see detail well below 1 millimeter, approaching the scale of neurons themselves. The development of room-temperature superconductors will soon allow brain imaging without the need for massive magnets. We'll be able to scan people's brains while they walk around a room, and then we'll be able to do the same thing with animals in a more natural environment. Optogenetics, which can turn neurons on and off with light, has already revolutionized our understanding of how specific circuits affect behavior, at least in mice.

Science has tremendous power to shape our moral intuitions. The knowledge gained about brain function can only increase our awareness of what it is like to be an animal. There will be lots of things in common with humans, and there will be differences. There is already ample evidence that many animals are aware of sensations and their environment. This is sentience. We do not yet know the full extent to which animals are self-aware, but I believe neuroscience will soon answer that question. Before then, people will need to decide how much consciousness—mere sentience or self-awareness—is necessary to afford legal consideration. Only then will people be able to advocate effectively for animals.

There is one final reason to care about how we treat animals and whether we care simply about suffering, or care about sentience as well. We, *Homo sapiens*, might soon be an animal in the eyes of our successors.

Natural selection is winding down for the human species. We are now in the beginning stages of a new form of evolution, if it can even be called that. Yuval Harari has called it the era of intelligent design, but there hasn't been much evidence of intelligence in what we have been doing to our genome or our environment. I would call it the era of tinkering.

In the tinkering era, the ease with which we can manipulate the genome has accelerated at a remarkable pace. A human genome was completely sequenced in 2003, after a decade-long effort and a cost of nearly $5 billion (in 2016 dollars). By 2016, the cost to generate a high-quality sequence for a single person had fallen to about $1,000.

But the most fearsome new technology is CRISPR/Cas9,

which was first described in 2012 as a tool to edit DNA in living cells. Cas9 is a protein that can unzip and cut a portion of DNA. It was first discovered in bacteria as a defense mechanism against invading viral DNA. But Cas9 can be given any short sequence of DNA to seek out and destroy, not just viruses. It has been used to cut out mutations, and once a cut is made, scientists can insert new DNA of their choosing in its place. More primitive DNA modification has already led to the creation of various types of hybrid animals. On the relatively harmless end of the spectrum, genetic engineers have designed fish and cats that can glow in the dark because they have a gene from a fluorescent jellyfish. With CRISPR, things are moving toward the creepier end, with genetic engineers putting snippets of human DNA into pigs so the pigs can grow organs ready for human transplantation. We may soon reach the point where it becomes difficult to define a species. How much human DNA would it take to make a pig not a pig?

Because these tinkerings alter the very DNA that make up cells, it is possible for the alterations to be passed on to offspring. So far, researchers have been careful to take measures so that these animals don't reproduce, but it is only a matter of time before an accidental mating happens. You can be sure of one thing: our underestimation of our ability to contain technology.

But I am not too worried about pig-human chimeras, or even future dogs with human genes for speech. I worry about the end of *Homo sapiens*, which is happening faster than most people realize. George Church, one of the leading proponents of genetic engineering, has said that it would be unethical *not* to fix the DNA of someone with cancer. It is hard to argue against that stance. And once we walk down the path of fixing "bad" DNA, it will be a short hop to improving "normal" DNA.

Humans have always had the urge to improve themselves. It is unreasonable to expect that prohibitions against tinkering with human embryos, like the ones the NIH has put in place, will be effective. Those with money will do it anyway, probably in some other country. Sexual reproduction, with all its messy DNA mergings, might become obsolete—at least for making children. And that is when a new species will come into existence.

Call it *Homo hominis*. Man of men.

Neanderthals coexisted for tens of thousands of years with *Homo sapiens*, but I don't expect we would last that long with *Homo hominis*. Maybe *H. hominis* would treat the planet better than we have. But *H. hominis* would be as far beyond us as we are beyond chimpanzees. We would be doing the future remnant of our species a favor by considering now what it means to be sentient and what rights that confers. Later will be the wrong time to ask if *sapiens* deserves to coexist with *hominis*, or whether future *sapiens* should be relegated to zoos.

Epilogue

The Brain Ark

Of course we didn't stop with the thylacine and the devil.
DTI could be used on the brain of any animal, and more
and more brains began to find their way to the lab. Mostly this
was the result of Peter's networking, but it became apparent that
there are a lot of exotic brains sitting on researchers' shelves and
in the vaults of museum collections. This was a potential gold-
mine for piecing together the puzzle of brain evolution.

The Marine Mammal Center sent us brains of other pinnipeds besides sea lions. Soon we had DTI data on harbor seals
and elephant seals. Peter even located the brains of two manatees in Florida. Although all of these marine mammals share
the same basic body morphology, they hunt for food in different
ways and have wildly different social lives. Harbor seals have
more vocal flexibility than sea lions, with the most famous
being Hoover, a harbor seal rescued and raised in Maine in the
1970s. Hoover became famous for yelling, "Hey! Hey! Come on
over here!" in a gruff Maine accent. And while sea lions don't
have a great range of vocalizations, elephant seals are famous for

the males' mating competition with intimidating vocalizations and fights. And manatees, well, don't do much of anything. These differences in something as simple as vocalizations, even in closely related species, have to be manifest in their brains, just as we found with the thylacine and the devil. These differences will help us understand what it is like to be a sea lion, a seal, or a manatee.

On the canid front, we started getting the brains of euthanized coyotes from the US Department of Agriculture's Predator Facility in Utah, which manages the largest captive colony of coyotes in North America. I have mixed feelings about such facilities, but coyotes have become ubiquitous even in urban environments, and understanding their psychology and behavior is necessary if we are to achieve a peaceful coexistence. In the Atlanta suburb where I live, one neighborhood has been pushing the city council to permit coyote hunting within the city limits. Of course, I am against such efforts. I enjoy the coyotes' plaintive yips and howls in the night, and I have high hopes that the brains we scanned will shed light on what makes coyotes tick and how they're different from dogs.

Even with this growing collection of brains, we are only scratching the surface. Almost all neuroscience research is currently concentrated in a handful of species. Research on human brains is foundational to understanding diseases like Alzheimer's and Parkinson's, as well as mental illnesses like schizophrenia and depression. Research on nonhuman primates, mostly monkeys, was, until recently, a major area of neuroscience research, but the growing recognition of primate sentience has led to a rapid decline in federal funding for that type of work. At the other end of the spectrum, research on rats and mice has exploded. These species are a key component of the Federal Brain

Research through Advancing Innovative Neurotechnologies (BRAIN) initiative, which seeks to fund the development of new technologies—such as optogenetics—that will ultimately change our understanding of the human brain.

But there are a lot of animals between rats and humans. Even if we're just talking about mammals, there are about 5,000. Why aren't we studying their brains, too?

The question becomes especially urgent in the face of growing evidence that many of the world's species are disappearing at an alarming rate. *Living Planet Report 2016*, published by the World Wildlife Fund (WWF), paints a grim picture. Loss of habitats, overfishing, pollution, the rise of invasive species, and climate change are all contributing to what many believe is the sixth great extinction event. These problems seem overwhelming, and it is easy to despair that they are unsolvable without the political and economic will necessary to confront them.

But we have to try.

I strongly share the WWF's perspective that we are "One Planet." The Earth is not a zero-sum game in which humanity has to survive at the expense of everything else. It requires only a small shift in perspective to see how the welfare of all species is tied together. The continuation of current unsustainable practices will, without a doubt, result in the extinction of many species. Ultimately, though, the human species will suffer, too, from altered weather patterns, rising sea levels, and emergent diseases.

My small part is to raise awareness of the mental lives of the animals with whom we share the planet.

As a start, my colleagues and I have launched the Brain Ark (http://brainark.org). Our goal is to catalog and describe the brains of the Earth's megafauna before they are gone. The Ark will contain three-dimensional reconstructions of the white matter pathways of each species, which can be linked to behavioral observations or ecological analyses. The data will be in sufficient detail that we can virtually probe how brain regions are connected to each other and therefore answer questions about brain evolution, or how brain structure is related to species-specific attributes, such as predator-prey relationships, their ecological niches, and foraging strategies.

The WWF estimates that two-thirds of many species' populations may be gone by 2020. Apart from the ecological catastrophe, scientific opportunities may be lost forever. It is imperative that we begin the archival process for all species, and especially for megafauna, so that we can preserve information about their brains at the highest fidelity possible with current technology. Such information may also help conservation efforts by determining how animals adapt to specific environments, especially in the face of habitat loss.

It may even help advocate for the animals who can't speak for themselves.

Acknowledgments

First, the dogs. What began with two dogs in 2011 has grown to more than thirty. Without them, and the passion of their owners, none of this would have been possible. Thanks to Barrington (Bob Weber), Bhubo (Ashwin Sakhardande), Dixie (Alexandria Andrews), Eli (Lindsay Fetters), Gemini (Sami Griffith), Huxley (Melanie Pincus), King Tubby (Leah Dawson), Koda and Zula (Cathy Siler), Mason (Chris McNamara), Mauja (Rebecca Beasley), McKenzie (Melissa Cate), Myrtle (Carrol Farren), Nelson (Jeff Petermann), Ninja (Sairina Merino Tsui), Nook (Van Nguyen), Obi (Liz Diaz), Oliver (Yusuf Uddin), Sierra (Diana Bush), Sophie (Rachel Purcell), Tigger (Aliza Levenson), and Wil (Emily Chapman).

Some dogs have been exceptional, participating in multiple experiments, so special thanks to Caylin (Lorrie Backer), Edmond (Marianne Ferraro), Jack (Cindy Keen), Kady (Patricia King), Libby (Claire Pearce Mancebo), Ohana (Cecilia Kurland), Ozzie (Patti Rudi), Pearl (Vicki D'Amico), Stella (Nicole Zitron), Tallulah (Anna and Cory Inman), Truffles (Diana Delatour), Tug (Jessa Fagan), Velcro (Lisa Tallant), and Zen (Darlene Coyne).

On the human side of the Dog Project, Mark Spivak has been a steady guide to training the dogs as well as a good friend and business partner in our efforts to use neuroscience to improve the training of service dogs at Dog Star Technologies. Marian Scopa, in addition to raising puppies, keeps us organized. Along with Mark, Andrew Brooks has been with the project since its inception. Andrew did the initial dog experiments and worked out much of the analysis pipeline, which we continue to use. Ashley Prichard is the latest to join the lab, but she has already contributed her own ideas for furthering the Dog Project. She did the word-object experiment.

I owe special thanks to Peter Cook. More than anyone else, he was responsible for opening my eyes to the mental lives of other animals. It was Peter who proposed using DTI in other species, and that idea changed the trajectory of research in the lab. Peter is one of those unique individuals who is knowledgeable about a wide range of topics besides science, and this informs everything he does. He, too, has moved on in his own academic career, and I will miss our philosophical conversations about animal minds. But I will still seek out guidance from his moral compass when mine is adrift.

Many people have made the DTI projects possible. Karla Miller shared her MRI sequences. Frances Gulland arranged for the shipment of pinniped brains, while Colleen Reichmuth shared her knowledge of pinniped behavior. Both were gracious hosts when I visited. Lori Marino pointed me to her collection of cetacean brains. For the marsupials, I am grateful to Esther Langan and Darrin Lunde at the Smithsonian Institution, Ken Ashwell at the University of New South Wales (UNSW), Marco Gruwel at the Biological Resources Imaging Laboratory at UNSW, Sandy Ingleby at the Australian Museum, Michael Archer at UNSW, Stephen Sleightholme, Col Bailey, Nicholas Clement, Kathryn Medlock at the Tasmanian Museum and Art

Gallery, and Carolyn Hogg and the Save the Tasmanian Devil Project.

A few people have been patient enough to read early drafts of chapters. It is a thankless task, and to do it well requires tact and an appreciation of writers' fragile egos. Thanks to Peter Cook, Lori Marino, Julia Haas, and Kathleen Berns. James Levine, at Levine-Greenberg-Rostan Literary Agency, has been a sage guide and strong advocate for this project. He matched me with TJ Kelleher at Basic Books. TJ has been the best editor I could have hoped for, pushing me where I needed to be pushed.

Finally, as always, special thanks to my wife, Kathleen, and children, Helen and Maddy, for their patience and support through yet another "last book I ever write." Helen took many of the photographs for the Dog Project, and Maddy made the website for the Brain Ark.

Notes

INTRODUCTION

3 **on to something important:** G. Berns, *How Dogs Love Us: A Neuroscientist and His Adopted Dog Decode the Canine Brain* (New York: New Harvest, 2013).

CHAPTER 1: WHAT IT'S LIKE TO BE A DOG

12 **"What Is It Like to Be a Bat?":** T. Nagel, "What Is It Like to Be a Bat?" *Philosophical Review* 83 (1974): 435–450.

14 **not the same thing as the experience itself:** The inner/outer dichotomy of experience had been questioned long before Nagel's essay. See L. Wittgenstein, *Philosophical Investigations*, translated by G. E. M. Anscombe, P. M. S. Hacker, and J. Schulte, 4th ed. (West Sussex, UK: Wiley-Blackwell, 2009).

14 **commonalities extend to other animals:** P. M. Churchland, "Some Reductive Strategies in Cognitive Neurobiology," *Mind* 95 (1986): 279–309; P. Godfrey-Smith, "On Being an Octopus," *Boston Review*, May/June 2013, 46–60.

15 **constitutes mental experience:** The astute reader will realize that the brain is a nonlinear system, and there is no guarantee that it is simply the sum of its parts. I think of the domains like photographic perspectives. A photograph is a two-dimensional representation of something. While a single photograph is insufficient to fully represent an object, if you had pictures from enough vantage points, you could reproduce the thing with high fidelity.

Mental domains may be snapshots of the mind from different vantage points.

15 **avoid something painful:** J. E. LeDoux, "Coming to Terms with Fear," *Proceedings of the National Academy of Sciences of the United States of America* 111 (2014): 2871–2878.

17 **caused severe brain trauma:** A. M. Owen, M. R. Coleman, M. Boly, M. H. Davis, S. Laureys, and J. D. Pickard, "Detecting Awareness in the Vegetative State," *Science* 313 (2006): 1402.

17 **remembering an image instead of seeing it:** K. N. Kay, T. Naselaris, R. J. Prenger, and J. L. Gallant, "Identifying Natural Images from Human Brain Activity," *Nature* 452 (2008): 352–356; S. Nishimoto, A. T. Vu, T. Naselaris, Y. Benjamini, B. Yu, and J. L. Gallant, "Reconstructing Visual Experiences from Brain Activity Evoked by Natural Movies," *Current Biology* 21 (2011): 1641–1646; T. Naselaris, C. A. Olman, D. E. Stansbury, K. Ugurbil, and J. L. Gallant, "A Voxel-Wise Encoding Model for Early Visual Areas Decodes Mental Images of Remembered Scenes," *NeuroImage* 105 (2015): 215–228.

CHAPTER 2: THE MARSHMALLOW TEST

38 **primates during similar tasks:** K. Rubia, S. Overmeyer, E. Taylor, M. Brammer, S. C. R. Williams, A. Simmons, C. Andrew, and E. T. Bullmore, "Functional Frontalisation with Age: Mapping Neurodevelopmental Trajectories with FMRI," *Neuroscience Biobehavioral Reviews* 24 (2000): 13–19; A. R. Aron, T. E. Behrens, S. Smith, M. J. Frank, and R. A. Poldrack, "Triangulating a Cognitive Control Network Using Diffusion-Weighted Magnetic Resonance Imaging (MRI) and Functional MRI," *Journal of Neuroscience* 27 (2007): 3743–3752; A. Aron, T. W. Robbins, and R. A. Poldrack, "Inhibition and the Right Inferior Frontal Cortex," *Trends in Cognitive Sciences* 8 (2004): 170–177.

39 **psychologist Walter Mischel:** W. Mischel, Y. Shoda, and M. L. Rodriguez, "Delay of Gratification in Children," *Science* 244 (1989): 933–938.

40 **imaging studies of Mischel's subjects:** B. J. Casey, L. H. Somerville, I. H. Gotlib, O. Ayduk, N. T. Franklin, M. K. Askren,

J. Jonides, et al. "Behavioral and Neural Correlates of Delay of Gratification 40 Years Later," *Proceedings of the National Academy of Sciences of the United States of America* 108 (2011): 14998–15003.

43 **immaturity of the frontal lobes:** B. Milner, "Effects of Different Brain Lesions on Card Sorting: The Role of the Frontal Lobes," *Archives of Neurology* 9 (1963): 90–100; A. M. Owen, A. C. Roberts, J. R. Hodges, B. A. Summers, C. E. Polkey, and T. W. Robbins, "Contrasting Mechanisms of Impaired Attentional Set-Shifting in Patients with Frontal Lobe Damage or Parkinson's Disease," *Brain* 116 (1993): 1159–1175.

43 **fail just like a nine-month-old infant:** A. Diamond and P. S. Goldman-Rakic, "Comparison of Human Infants and Rhesus Monkeys on Piaget's AB Task: Evidence for Dependence on Dorsolateral Prefrontal Cortex," *Experimental Brain Research* 74 (1989): 24–40.

43 **monkeys, apes, lemurs, birds, elephants, rodents, and dogs:** E. L. MacLean, B. Hare, C. L. Nunn, E. Addessi, F. Amici, R. C. Anderson, F. Aureli, et al., "The Evolution of Self-Control," *Proceedings of the National Academy of Sciences of the United States of America* 111 (2014): E2140–E2148.

Chapter 3: Why a Brain?

51 **next fifty years:** W. James, *The Principles of Psychology* (New York: Henry Holt, 1890); I. P. Pavlov, *Conditioned Reflexes* (Oxford: Oxford University Press, 1927); E. L. Thorndike, *Animal Intelligence* (New York: Macmillan, 1911); B. F. Skinner, *The Behavior of Organisms: An Experimental Analysis* (New York: Appleton-Century-Crofts, 1938).

52 **information was manipulated:** A. Newell and H. A. Simon, *Human Problem Solving* (New York: Prentice-Hall, 1972).

53 **godlike hand of a programmer:** D. E. Rumelhart, J. L. McClelland, and PDP Research Group, *Parallel Distributed Processing: Explorations in the Microstructure of Cognition* (Cambridge, MA: MIT Press, 1986); P. S. Churchland and T. J. Sejnowski, *The Computational Brain* (Cambridge, MA: MIT Press, 1992).

54 **animals with nervous systems have muscles:** G. Jékeley, F. Keijzer, and P. Godfrey-Smith, "An Option Space for Early Neural Evolution," *Philosophical Transactions of the Royal Society B* 370 (2015): 20150181.

54 **environment in which it is situated:** F. J. Varela, E. Thompson, and E. Rosch, *The Embodied Mind: Cognitive Science and Human Experience* (Cambridge, MA: MIT Press, 1991).

54 **active perception:** L. P. J. Selen, M. N. Shadlen, and D. M. Wolpert, "Deliberation in the Motor System: Reflex Gains Track Evolving Evidence Leading to a Decision," *Journal of Neuroscience* 32 (2012): 2276–2286.

54 **tight relationship:** A. R. Damasio, *Descartes' Error: Emotion, Reason, and the Human Brain* (New York: G. P. Putnam, 1994).

59 **more complex body shapes:** N. Shubin, *Your Inner Fish: A Journey into the 3.5-Billion-Year History of the Human Body* (New York: Pantheon, 2008).

60 **laid eggs like reptiles and birds:** M. Ruta, J. Botha-Brink, S. A. Mitchell, and M. J. Benton, "The Radiation of Cynodonts and the Ground Plan of Mammalian Morphological Diversity," *Proceedings of the Royal Society of London B* 280 (2013): 20131865.

61 **intense debate for over one hundred years:** G. von Bonin, "Brain-Weight and Body-Weight of Mammals," *Journal of General Psychology* 16 (1937): 379–389; K. S. Lashley, "Persistent Problems in the Evolution of Mind," *Quarterly Review of Biology* 24 (1949): 28–42; L. Chittka and J. Niven, "Are Bigger Brains Better?" *Current Biology* 19 (2009): R995–R1008.

61 **". . . performing the function":** H. J. Jerison, *Evolution of the Brain and Intelligence* (New York: Academic, 1973).

62 **raised to the exponent 2/3:** G. Roth and U. Dicke, "Evolution of the Brain and Intelligence," *Trends in Cognitive Sciences* 9 (2005): 250–257.

62 **volume raised to the 2/3 power:** For example, the area of a sphere is proportional to r^2, where r is the radius, while the volume is proportional to r^3. Therefore, the area is proportional to the volume$^{2/3}$.

62 **always exceptions:** T. W. Deacon, "Rethinking Mammalian Brain Evolution," *American Zoologist* 30 (1990): 629–705.

63 **makes a person smarter:** While there is good evidence that diet and exercise have positive effects on cognitive function, this is likely due to the release of neural growth factors rather than changes in body mass.

63 **number of neurons in a brain:** S. Herculano-Houzel, *The Human Advantage: A New Understanding of How Our Brain Became Remarkable* (Cambridge, MA: MIT Press, 2016).

65 **coordinated fashion:** B. L. Finlay and R. B. Darlington, "Linked Regularities in the Development and Evolution of Mammalian Brains," *Science* 268 (1995): 1578–1584.

66 **evolutionary pressures:** R. A. Barton and P. H. Harvey, "Mosaic Evolution of Brain Structure in Mammals," *Nature Neuroscience* 405 (2000): 1055–1058.

66 **top of their brains:** H. J. Karten, "Vertebrate Brains and Evolutionary Connectomics: On the Origins of the Mammalian 'Neocortex,'" *Proceedings of the Royal Society of London B* 370 (2015): 20150060.

67 **larger hippocampi than those that do not:** J. R. Krebs, D. F. Sherry, S. D. Healy, V. H. Perry, and A. L. Vaccarino, "Hippocampal Specialization of Food-Storing Birds," *Proceedings of the National Academy of Sciences of the United States of America* 86 (1989): 1388–1392.

68 **powerful relationship between white and gray matter:** K. Zhang and T. J. Sejnowski, "A Universal Scaling Law Between Gray Matter and White Matter of Cerebral Cortex," *Proceedings of the National Academy of Sciences of the United States of America* 97 (2000): 5621–5626.

71 **study of connectomics:** S. Seung, *Connectome: How the Brain's Wiring Makes Us Who We Are* (New York: Houghton Mifflin Harcourt, 2012).

72 **coordinated electrical activity:** S. Dehaene, *Consciousness and the Brain: Deciphering How the Brain Codes Our Thoughts* (New York: Penguin, 2014).

CHAPTER 4: SEIZING SEA LIONS

75 **ability to form new memories:** T. M. Perl, L. Bedard, T. Kosatsky, E. C. D. Todd, and R. S. Remis, "An Outbreak of Toxic

Encephalopathy Caused by Eating Mussels Contaminated with Domoic Acid," *New England Journal of Medicine* 322 (1990): 1775–1780.

76 **burns up the surrounding cells:** H. Parfenova, S. Basuroy, S. Bhattacharya, D. Tcheranova, Y. Qu, R. F. Regan, and C. W. Leffler, "Glutamate Induces Oxidative Stress and Apoptosis in Cerebral Vascular Endothelial Cells: Contributions of HO-1 and HO-2 to Cytoprotection," *American Journal of Physiology—Cell Physiology* 290 (2006): C1399–C1410.

76 **including the hippocampus and the amygdala:** R. S. Teitelbaum, R. J. Zatorre, S. Carpenter, D. Gendron, A. C. Evans, and A. Gjedde, "Neurologic Sequelae of Domoic Acid Intoxication Due to the Ingestion of Contaminated Mussels," *New England Journal of Medicine* 322 (1990): 1781–1787.

76 **fanned out over Prince Edward Island:** S. S. Bates, C. J. Bird, A. S. W. de Freitas, R. Foxall, M. Gilgan, L. A. Hanic, G. R. Johnson, et al., "Pennate Diatom *Nitzschia Pungens* as the Primary Source of Domoic Acid, a Toxin in Shellfish from Eastern Prince Edward Island, Canada," *Canadian Journal of Fisheries and Aquatic Sciences* 46 (1989): 1203–1215.

80 **findings were a bombshell:** C. A. Scholin, F. Gulland, G. J. Doucette, S. Benson, M. Busman, F. P. Chavez, J. Cordaro, et al., "Mortality of Sea Lions Along the Central California Coast Linked to a Toxic Diatom Bloom," *Nature* 403 (2000): 80–84.

90 **errors a sea lion made:** P. F. Cook, C. Reichmuth, A. A. Rouse, L. A. Libby, S. E. Dennison, O. T. Carmichael, K. T. Kruse-Elliott, et al., "Algal Toxin Impairs Sea Lion Memory and Hippocampal Connectivity, with Implications for Strandings," *Science* 350 (2015): 1545–1547.

91 **exactly what it feels like:** "What Is Epilepsy?," Epilepsy Foundation, n.d., www.epilepsy.com/learn/epilepsy-101/what-epilepsy, retrieved March 16, 2016.

91 **vivid accounts of his fits:** M. Costandi, "Diagnosing Dostoyevsky's Epilepsy," 2007, https://neurophilosophy.wordpress.com/2007/04/16/diagnosing-dostoyevskys-epilepsy, retrieved March 16, 2016.

91 " . . . semi-dazed condition": F. Dostoyevsky, *The Idiot*, 1868, Project Gutenberg, 2012.

98 downstream effect is an increase in connectivity: L. Bonilha, T. Nesland, G. U. Martz, J. E. Joseph, M. V. Spampinato, J. C. Edwards, and A. Tabesh, "Medial Temporal Lobe Epilepsy Is Associated with Neuronal Fibre Loss and Paradoxical Increase in Structural Connectivity of Limbic Structures," *Journal of Neurology, Neurosurgery and Psychiatry* 83, no. 9 (2012); V. Dinkelacker, R. Valabregue, L. Thivard, S. Lehéricy, M. Baulac, S. Samson, and S. Dupont, "Hippocampal⬚Thalamic Wiring in Medial Temporal Lobe Epilepsy: Enhanced Connectivity Per Hippocampal Voxel," *Epilepsia* 56 (2015): 1217–1226.

CHAPTER 5: RUDIMENTS

103 gave rise to hominids: S. Pinker, *The Language Instinct* (New York: William Morrow, 1994).

104 understand simple phrases: R. J. Schusterman and K. Krieger, "California Sea Lions Are Capable of Semantic Representation," *Psychological Record* 34 (1984): 3–23.

104 understanding pictorial symbols: R. J. Schusterman, C. R. Kastak, and D. Kastak, "The Cognitive Sea Lion: Meaning and Memory in the Laboratory and in Nature," in *The Cognitive Animal: Empirical and Theoretical Perspectives on Animal Cognition*, edited by M. Bekoff, C. Allen, and G. M. Burghardt, 217–228 (Cambridge, MA: MIT Press, 2002).

105 when spoken, has a rhythm: There are many other aspects to language as well, such as grammar, syntax, and recursion.

105 must have its origin in other species: P. Kivy, "Charles Darwin on Music," *Journal of the American Musicological Society* 12 (1959): 42–48.

106 " . . . nature of their nervous systems": C. Darwin, *The Descent of Man, and Selection in Relation to Sex* (London: John Murray, 1871).

106 vocal mimic theory of rhythm: A. D. Patel, "Musical Rhythm, Linguistic Rhythm, and Human Evolution," *Music Perception: An Interdisciplinary Journal* 24 (2006): 99–104.

109 **"rate-flexibility":** P. Cook, A. Rouse, M. Wilson, and C. Reichmuth, "A California Sea Lion (*Zalophus Californianus*) Can Keep the Beat: Motor Entrainment to Rhythmic Auditory Stimuli in a Non Vocal Mimic," *Journal of Comparative Psychology* 127 (2013): 412.

111 **band should be renamed "Earth, Wind, Fire, and Water":** See "Sea Lion Dances to 'Boogie Wonderland,'" YouTube, posted April 2, 2013, https://www.youtube.com/watch?v=KUfRS m8NTZg; "Beat Keeping in a California Sea Lion," YouTube, posted March 31, 2013, https://www.youtube.com/watch ?v=6yS6qU_w3JQ.

112 **simple, steady rhythm:** M. Dhamala, G. Pagnoni, K. Wiesenfeld, C. F. Zink, M. Martin, and G. S. Berns, "Neural Correlates of the Complexity of Rhythmic Finger Tapping," *NeuroImage* 20 (2003): 918–926.

112 **and even autism:** S. H. Fatemi, K. A. Aldinger, P. Ashwood, M. L. Bauman, C. D. Blaha, G. J. Blatt, A. Chauhan, et al., "Consensus Paper: Pathological Role of the Cerebellum in Autism," *The Cerebellum* 11 (2012): 777–807.

115 **same dynamics in a sea lion as in a human:** A. A. Rouse, P. F. Cook, E. W. Large, and C. Reichmuth, "Beat Keeping in a Sea Lion as Coupled Oscillation: Implications for Comparative Understanding of Human Rhythm," *Frontiers in Neuroscience* 10, no. 257 (2016).

CHAPTER 6: PAINTING WITH SOUND

117 **modern understanding of dolphin brain anatomy:** L. Marino, T. L. Murphy, A. L. DeWeerd, J. A. Morris, S. H. Ridgway, A. J. Fobbs, N. Humblot, and J. I. Johnson, "Anatomy and Three-Dimensional Reconstructions of the Brain of a White Whale (*Delphinapterus Leucas*) from Magnetic Resonance Images," *Anatomical Record* 262 (2001): 429–439; L. Marino, K. D. Sudheimer, D. A. Pabst, W. A. McLellan, D. Filsoof, and J. I. Johnson, "Neuroanatomy of the Common Dolphin (*Delphinus Delphis*) as Revealed by Magnetic Resonance Imaging," *Anatomical Record* 268 (2002): 411–429.

118 **than if they did not have the marks:** D. Reiss and L. Marino, "Mirror Self-Recognition in the Bottlenose Dolphin: A Case of Cognitive Convergence," *Proceedings of the National Academy of Sciences of the United States of America* 98 (2001): 5937–5942.

122 **10 percent of the brain could be called frontal:** H. H. A. Oelschlager and J. S. Oelschlager, "Brain," in *Encyclopedia of Marine Mammals*, edited by W. F. Perrin, B. Wursig, and J. G. M. Thewissen, 134–149 (Burlington, MA: Academic Press, 2009).

127 **"monkey lips" because of their shape:** A. S. Frankel, "Sound Production," in *Encyclopedia of Marine Mammals*, edited by W. F. Perrin, B. Wursig, and J. G. M. Thewissen, 1056–1071 (Burlington, MA: Academic Press, 2009).

128 **differed by as little as 0.3 millimeters:** W. W. L. Au, "Echolocation," in *Encyclopedia of Marine Mammals*, edited by W. F. Perrin, B. Wursig, and J. G. M. Thewissen, 348–357 (Burlington, MA: Academic Press, 2009).

131 **substances that tracked down axons:** V. V. Popov, T. F. Ladygina, and A. Y. Supin, "Evoked Potentials of the Auditory Cortex of the Porpoise, *Phocoena Phocoena*," *Journal of Comparative Physiology* A 158 (1986): 705–711; V. E. Sokolov, T. F. Ladygina, and A. Y. Supin, "Localization of Sensory Zones in the Dolphin's Cerebral Cortex," *Doklady Akademy Nauk SSSR* 202 (1972): 490–493; A. V. Revishchin and L. J. Garey, "The Thalamic Projection to the Sensory Neocortex of the Porpoise, *Phocoena Phocoena*," *Journal of Anatomy* 169 (1990): 85–102.

131 **visual information in terrestrial mammals:** Oelschlager and Oelschlager, "Brain."

134 **translates to a distance map:** M. Kossl, J. C. Hechavarria, C. Voss, S. Macias, E. C. Mora, and M. Vater, "Neural Maps for Target Range in the Auditory Cortex of Echolating Bats," *Current Opinion in Neurobiology* 24 (2014): 68–75.

135 **separated by 80 million years:** J. Parker, G. Tsagkogeorga, J. A. Cotton, Y. Liu, P. Provero, E. Stupka, and S. J. Rossiter, "Genome-Wide Signatures of Convergent Evolution in Echolocating Mammals," *Nature* 502 (2013): 228–231.

CHAPTER 7: BURIDAN'S ASS

138 **lurid headline:** M. Wells, "In Search of the Buy Button," *Forbes*, September 1, 2003, 62–70.

139 **circulated a paper:** Vul eventually had to retitle the paper to something more prosaic: see E. Vul, C. Harris, P. Winkielman, and H. Pashler, "Puzzlingly High Correlations in fMRI Studies of Emotion, Personality, and Social Cognition," *Perspectives on Psychological Science* 4 (2009): 274–290.

140 **deducing mental processes from brain activity:** R. A. Poldrack, "Can Cognitive Processes Be Inferred from Neuroimaging Data?," *Trends in Cognitive Sciences* 10 (2006): 59–63.

141 **or even a mind:** L. Barrett, "Why Brains Are Not Computers, Why Behaviorism Is Not Satanism, and Why Dolphins Are Not Aquatic Apes," *Behavior Analyst* 39 (2016): 9–23.

141 **" . . . dogs' internal states":** G. Berns, "Dogs Are People, Too," *New York Times*, October 5, 2013.

143 **Think dentist:** G. S. Berns, J. C. Chappelow, M. Cekic, C. F. Zink, G. Pagnoni, and M. E. Martin-Skurski, "Neurobiological Substrates of Dread," *Science* 312 (2006): 754–758.

153 **demonstrated in pigeons:** R. J. Herrnstein, "Relative and Absolute Strength of Response as a Function of Frequency of Reinforcement," *Journal of the Experimental Analysis of Behavior* 4 (1961): 267–272.

153 **dies of hunger and thirst:** The paradox actually dates to Aristotle in *On the Heavens*.

153 **frozen in indecision:** M. Hauskeller, "Why Buridan's Ass Doesn't Starve," *Philosophy Now* 81 (2010): 9.

156 **regretting something in the future:** G. Loomes and R. Sugden, "Regret Theory: An Alternative Theory of Rational Choice Under Uncertainty," *Economic Journal* 92 (1982): 805–824.

156 **no longer experienced it:** N. Camille, G. Coricelli, J. Sallet, P. Pradat-Diehl, J.-R. Duhamel, and A. Sirigu, "The Involvement of the Orbitofrontal Cortex in the Experience of Regret," *Science* 304 (2004): 1167–1170.

156 **neural basis of regret in rats:** A. P. Steiner and A. D. Redish, "Behavioral and Neurophysiological Correlates of Regret in Rat Decision-Making on a Neuroeconomic Task," *Nature Neuroscience* 17 (2014): 995–1002.

CHAPTER 8: TALK TO THE ANIMALS

160 **two hundred words for different objects:** J. Kaminski, J. Call, and J. Fischer, "Word Learning in a Domestic Dog: Evidence for 'Fast Mapping,'" *Science* 304 (2004): 1682–1683.

160 **more than one thousand words:** J. W. Pilley and A. K. Reid, "Border Collie Comprehends Object Names as Verbal Referents," *Behavioural Processes* 86 (2011): 1641–1646.

162 **based solely on his or her brain activity:** S. Nishimoto, A. T. Vu, T. Naselaris, Y. Benjamini, B. Yu, and J. L. Gallant, "Reconstructing Visual Experiences from Brain Activity Evoked by Natural Movies," *Current Biology* 21 (2011): 1641–1646.

162 **Gallant's lab took a similar approach with language:** A. G. Huth, W. A. de Heer, T. L. Griffiths, F. E. Theunissen, and J. L. Gallant, "Natural Speech Reveals the Semantic Maps That Tile Human Cerebral Cortex," *Nature* 532 (2016): 453–458.

165 **generalized from learned words:** E. Van der Zee, H. Zulch, and D. Mills, "Word Generalization by a Dog (*Canis Familiaris*): Is Shape Important?," *PLoS ONE* 7, no. 11 (2012): e49382.

165 **appears around age two:** B. Landau, L. B. Smith, and S. S. Jones, "The Importance of Shape in Early Lexical Learning," *Cognitive Development* 3 (1988): 299–321.

166 **cannot understand that words can refer to objects:** P. Bloom, "Can a Dog Learn a Word?," *Science* 304 (2004): 1605–1606.

166 **similar capacity during domestication:** S. Pinker and R. Jackendoff, "The Faculty of Language: What's Special About It?," *Cognition* 95 (2005): 201–236.

166 **modifiers for size and color:** R. J. Schusterman and K. Krieger, "California Sea Lions Are Capable of Semantic Representation," *Psychological Record* 34 (1984): 3–23.

167 **abilities in cetaceans:** J. C. Lilly, *Communication Between Man and Dolphin: The Possibilities of Talking with Other Species* (New York: Julian Press, 1978).

167 **rules of syntax were violated:** L. M. Herman, D. G. Richards, and J. P. Wolz, "Comprehension of Sentences by Bottlenosed Dolphins," *Cognition* 16 (1984): 129–219; L. M. Herman, S. A. Kuczaj, and M. D. Holder, "Responses to Anomalous Gestural Sequences by a Language-Trained Dolphin: Evidence for Processing of Semantic Relations and Syntactic Information," *Journal of Experimental Psychology: General* 122 (1993): 184–194.

169 **might look like graphically:** A. G. Huth, S. Nishimoto, A. T. Vu, and J. L. Gallant, "A Continuous Semantic Space Describes the Representation of Thousands of Object and Action Categories Across the Human Brain," *Neuron* 76 (2012): 1210–1224.

170 **faces they hadn't seen before:** C. A. Muller, K. Schmitt, A. L. A. Barber, and L. Huber, "Dogs Can Discriminate Emotional Expression of Human Faces," *Current Biology* 25 (2015): 1–5.

172 **compared to inanimate objects:** C. G. Gross, C. E. Rocha-Miranda, and D. B. Bender, "Visual Properties of Neurons in Inferotemporal Cortex of the Macaque," *Journal of Neurophysiology* 35 (1972): 96–111; R. Desimone, T. D. Albright, C. G. Gross, and C. Bruce, "Stimulus-Selective Properties of Inferior Temporal Neurons in the Macaque," *Journal of Neuroscience* 4 (1984): 2051–2062; D. Y. Tsao, S. Moeller, and W. A. Freiwald, "Comparing Face Patch Systems in Macaques and Humans," *Proceedings of the National Academy of Sciences of the United States of America* 105 (2008): 19514–19519.

174 **"dog face area," or DFA:** D. D. Dilks, P. A. Cook, S. K. Weiller, H. P. Berns, M. Spivak, and G. S. Berns, "Awake fMRI Reveals a Specialized Region in Dog Temporal Cortex for Face Processing," *PeerJ* 3 (2015): e1115.

174 **confirmed the finding a year later:** L. V. Cuaya, R. Hernández-Pérez, and L. Concha, "Our Faces in the Dog's Brain: Functional Imaging Reveals Temporal Cortex Activation During Perception of Human Faces," *PLoS ONE* 11, no. 3 (2016): e0149431.

174 **Sheep do:** K. M. Kendrick and B. A. Baldwin, "Cells in Temporal Cortex of Conscious Sheep Can Respond Preferentially to the Sight of Faces," *Science* 236 (1987): 448–450.

174 **Goats do:** C. Nawroth, J. M. Brett, and A. G. McElligott, "Goats Display Audience-Dependent Human-Directed Gazing Behaviour in a Problem-Solving Task," *Biology Letters* 12 (2016): 20160283.

174 **so do some birds:** J. M. Marzluff, R. Miyaoka, S. Minoshima, and D. J. Cross, "Brain Imaging Reveals Neuronal Circuitry Underlying the Crow's Perception of Human Faces," *Proceedings of the National Academy of Sciences of the United States of America* 109 (2012): 15912–15917.

175 **familiar and unfamiliar cows:** M. Coulon, B. L. Deputte, Y. Heyman, and C. Baudoin, "Individual Recognition in Domestic Cattle (*Bos Taurus*): Evidence from 2D Images of Heads from Different Breeds," *PLoS ONE* 4 (2009): e4441.

175 **their own faces in mirrors:** J. M. Plotnick, F. B. M. de Waal, and D. Reiss, "Self-Recognition in an Asian Elephant," *Proceedings of the National Academy of Sciences of the United States of America* 103 (2006): 17053–17057.

175 **maybe they just don't care:** S. G. Lomber and P. Cornwell, "Dogs, but Not Cats, Can Readily Recognize the Face of Their Handler," *Journal of Vision* 5 (2005): 49.

178 **mental effort to process:** T. Raettig and S. A. Kotz, "Auditory Processing of Different Types of Pseudo-Words: An Event-Related fMRI Study," *NeuroImage* 39 (2008): 1420–1428.

179 **verbs in the English language:** C. Fellbaum, "Wordnet and Wordnets," in *Encyclopedia of Language and Linguistics*, edited by K. Brown, 665–670 (Oxford: Elsevier, 2005).

179 **reason remains unclear:** S. Waxman, X. Fu, S. Arunachalam, E. Leddon, K. Geraghty, and H. Song, "Are Nouns Learned Before Verbs?," *Child Development Perspectives* 7 (2013): 155–159.

CHAPTER 9: A DEATH IN TASMANIA

183 **days were growing longer:** This story is reconstructed from the research of Robert Paddle. See R. Paddle, *The Last Tasmanian*

Tiger: The History and Extinction of the Thylacine (Cambridge: Cambridge University Press, 2000).

185 **still had the clipping:** "Beauty and the Beast at the Hobart Zoo: Girl Whose Greatest Chum Is a Full-Grown Leopard," *Register News-Pictorial*, May 17, 1930, http://trove.nla.gov.au/newspaper /article/54241041.

189 **" . . . forwarded to the Museum":** S. R. Sleightholme, "Confirmation of the Gender of the Last Captive Thylacine," *Australian Zoologist* 35 (2011): 953–956.

189 **film of captive thylacines:** "Tasmanian Tiger / Thylacine Combined Footage," YouTube, posted May 3, 2007, https://www.you tube.com/watch?v=odswge5onwY.

191 **competition when it finally arrived:** D. Quammen, *The Song of the Dodo: Island Biogeography in an Age of Extinctions* (New York: Scribner, 1996).

191 **4,650 years old:** E. Guiler and P. Godard, *Tasmanian Tiger: A Lesson to Be Learnt* (Perth, Australia: Abrolhos, 1998).

194 **first thylacine family:** C. Wemmer, "Opportunities Lost: Zoos and the Marsupial That Tried to Be a Wolf," *Zoo Biology* 21 (2002): 1–4.

195 **Apparently, nothing:** Ibid.

199 **all things related to the thylacine:** C. R. Campbell, "The Thylacine Museum," www.naturalworlds.org/thylacine/index.htm, retrieved April 19, 2016.

201 **reports of a "tyger":** E. R. Guiler, *Thylacine: The Tragedy of the Tasmanian Tiger* (Melbourne: Oxford University Press, 1985), 14.

201 **rid the island of them:** Ibid., 16.

202 **sanctuaries went unheeded:** Ibid., 138.

202 **they had given up:** Ibid., 140.

202 **" . . . no wool torn off the victim":** E. R. Guiler and G. K. Meldrum, "Suspected Sheep Killing by the Thylacine *Thylacinus cynocephalus* (Harris)," *Australian Journal of Science* 20 (1958): 214, reprinted in Guiler, *Thylacine*, 141.

207 **textbook on the topic:** K. W. S. Ashwell, ed., *The Neurobiology of Australian Marsupials* (Cambridge: Cambridge University Press, 2010).

208 **still in Australia:** C. Bailey, *Lure of the Thylacine: True Stories and Legendary Tales of the Tasmanian Tiger* (Victoria, Australia: Echo Publishing, 2016).

CHAPTER 10: LONESOME TIGER

215 **54 percent of these deaths:** US Department of Agriculture, "Sheep and Lamb Predator and Nonpredator Death Loss in the United States," 2015, USDA-APHIS-VS-CEAH-NAHMS.

215 **biomechanical structure for pursuit:** B. Figueirido and C. M. Janis, "The Predatory Behaviour of the Thylacine: Tasmanian Tiger or Marsupial Wolf?," *Biology Letters* (2011): 937–940.

215 **leads to similar conclusions:** M. E. Jones and D. M. Stoddart. "Reconstruction of the Predatory Behaviour of the Extinct Marsupial Thylacine (*Thylacinus cynocephalus*)," *Journal of Zoology* 246 (1998): 239–246.

217 **closest living relative to the thylacine:** W. Miller, D. I. Drautz, J. E. Janecka, A. M. Lesk, A. Ratan, L. P. Tomsho, M. Packard, et al., "The Mitochondrial Genome Sequence of the Tasmanian Tiger (*Thylacinus cynocephalus*)," *Genome Research* 19 (2009): 213–220.

218 **type of cell surrounding neurons:** E. P. Murchison, C. Tovar, A. Hsu, H. S. Bender, P. Kheradpour, C. A. Rebbeck, D. Obendorf, et al., "The Tasmanian Devil Transcriptome Reveals Schwann Cell Origins of a Clonally Transmissible Cancer," *Science* 327 (2010): 84–87.

218 **fatal like the devil cancer:** C. Murgia, J. K. Pritchard, S. Y. Kim, A. Fassati, and R. A. Weiss, "Clonal Origin and Evolution of a Transmissible Cancer," *Cell* 126 (2006): 477–487.

218 **declared endangered that year:** Save the Tasmanian Devil, www.tassiedevil.com.au/tasdevil.nsf, retrieved May 17, 2016.

219 **even then with mixed success:** T. D. Beeland, *The Secret World of Red Wolves: The Fight to Save North America's Other Wolf* (Chapel Hill: University of North Carolina Press, 2013).

219 **obtaining the brain of a Tasmanian devil for comparison:** The Smithsonian had also lent us the brain of a Tasmanian devil preserved from the same era as the thylacine, but the devil brain had shrunk, too.

221 **separate representations of motor and sensory information:** A. A. Abbie, "The Excitable Cortex in Perameles, Sarcophilus, Dasyurus, Trichosurus and Wallabia (Macropus)," *Journal of Comparative Neurology* 72 (1940): 469–487; L. Krubitzer, "The Magnificent Compromise: Cortical Field Evolution in Mammals," *Neuron* 56 (2007).

223 **emotion/motivation region:** G. S. Berns and K. W. S. Ashwell, "Reconstruction of the Cortical Maps of the Tasmanian Tiger and Comparison to the Tasmanian Devil," *PLoS ONE* 12 (2017): e0168993.

224 **any mammal that has ever lived:** S. Wroe, C. McHenry, and J. Thomason, "Bite Club: Comparative Bite Force in Big Biting Mammals and the Prediction of Predatory Behaviour in Fossil Taxa," *Proceedings of the Royal Society of London B* 272 (2005): 619–625.

225 **unknown outside of Australia:** This section is based on N. Clements, *The Black War: Fear, Sex, and Resistance in Tasmania* (Queensland, Australia: University of Queensland Press, 2014).

226 **Aboriginal girls and women:** Ibid., 17.

226 **height of absurdity:** Ibid., 44–45.

226 **" . . . hatred and fear":** Ibid., 49.

226 **arson and stock-killing:** Ibid., 60.

226 **killing thousands during the war:** Ibid., 89.

227 **paid for every male thylacine and seven for females:** E. R. Guiler, *Thylacine: The Tragedy of the Tasmanian Tiger* (Melbourne: Oxford University Press, 1985), 16.

227 **Soho district for a week:** Lee Jackson, "Victorian Money: How Much Did Things Cost?," Dictionary of Victorian London, www .victorianlondon.org/finance/money.htm, retrieved May 26, 2016.

228 **" . . . snuggle with dogs for warmth":** Personal email. May 6, 2016.

229 **near the Snake River:** C. Bailey, *Shadow of the Thylacine: One Man's Epic Search for the Tasmanian Tiger* (Victoria, Australia: Five Mile Press, 2013).

231 **printed out the directions:** The directions can be found at Tastracks, http://tastracks.webs.com/southwest.htm#529486938, retrieved June 2, 2016.

CHAPTER 11: DOG LAB

240 **finally stopped in 2016:** "Last Remaining Medical School to Use Live Animals for Training Makes Switch to Human-Relevant Methods," Physicians Committee for Responsible Medicine, June 30, 2016, https://www.pcrm.org/last_animal_lab, retrieved July 22, 2016.

240 **" . . . Can they *suffer*":** J. Bentham, *The Principles of Morals and Legislation* (Amherst: Prometheus Books, 1988).

241 **handled in research:** T. Cowan, "The Animal Welfare Act: Background and Selected Animal Welfare Legislation," Congressional Research Service, 2013.

241 **food or fiber:** United States Code, 2013 edition, Chapter 54, "Transportation, Sale, and Handling of Certain Animals," https://www.gpo.gov/fdsys/pkg/USCODE-2013-title7/html/USCODE-2013-title7-chap54.htm.

241 **replacement, reduction, and refinement:** W. M. S. Russell and R. L. Burch, *The Principles of Humane Experimental Technique* (London: Methuen, 1959).

242 **three R's were routinely ignored:** P. Singer, *Animal Liberation: The Definitive Classic of the Animal Movement* (New York: HarperCollins, 2009).

243 **" . . . substrates that generate consciousness":** "The Cambridge Declaration on Consciousness," Francis Crick Memorial Conference, July 7, 2012, http://fcmconference.org/img/CambridgeDeclarationOnConsciousness.pdf, retrieved August 3, 2016.

243 **767,622 animals were used in research:** "Annual Report Animal Usage by Fiscal Year," US Department of Agriculture, Animal and Plant Health Inspection Service, June 2016, https://www.aphis.usda.gov/animal_welfare/downloads/7023/Annual-Reports-FY2015.pdf, retrieved August 1, 2016.

244 **25 million to 100 million per year:** "Questions and Answers About Biomedical Research," Humane Society of the United States, n.d., www.humanesociety.org/issues/biomedical_research/qa/questions_answers.html; "Mice and Rats in Laboratories," People for the Ethical Treatment of Animals, n.d., www.peta.org/issues/animals-used-for-experimentation/animals-laboratories/mice-rats-laboratories, retrieved August 1, 2016.

245 **consistent with our treatment of animals:** H. Herzog, *Some We Love, Some We Hate, Some We Eat: Why It's So Hard to Think Straight About Animals* (New York: Harper, 2010).

246 **Carnegie Mellon University:** M. Botvinick and J. Cohen, "Rubber Hands 'Feel' Touch That Eyes See," *Nature* 391 (1998): 756.

247 **same illusion:** M. Wada, K. Takano, H. Ora, M. Ide, and K. Kansaku, "The Rubber Tail Illusion as Evidence of Body Ownership in Mice," *Journal of Neuroscience* 36 (2016): 11133–11137.

247 **people's moral sentiments:** J. Greene and J. Cohen, "For the Law, Neuroscience Changes Nothing and Everything," *Philosophical Transactions of the Royal Society B* 359 (2004): 1775–1785.

248 **Judge Matthew Cooper:** *Travis v. Murray*, Supreme Court New York County (NY Slip Op 23405, 42 Misc 3d 447), 2013.

249 **" . . . mere property":** *Rabideau v. City of Racine*, Supreme Court of Wisconsin (243 Wis 2d 486, 491, 627 NW2d, 795, 798), 2001.

253 **need for massive magnets:** A. Galante, R. Sinibaldi, A. Conti, C. De Luca, N. Catallo, P. Sebastiani, V. Pizzella, et al., "Fast Room Temperature Very Low Field-Magnetic Resonance Imaging System Compatible with Magnetoencephalography Environment," *PLoS ONE* 10 (2015): e0142701.

253 **our genome or our environment:** Y. N. Harari, *Sapiens: A Brief History of Humankind* (New York: HarperCollins, 2015).

254 **fallen to about $1,000:** "The Cost of Sequencing a Human Genome," National Human Genome Research Institute, updated July 6, 2016, https://www.genome.gov/27565109/the-cost-of-sequencing-a-human-genome, retrieved July 30, 2016.

255 **Man of men:** Yuval Harari calls the future species *Homo deus*, but I can't stomach calling anyone godlike. See Y. N. Harari, *Homo Deus: A Brief History of Tomorrow* (New York: HarperCollins, 2017).

EPILOGUE: THE BRAIN ARK

259 **grim picture:** *Living Planet Report 2016: Risk and Resilience in a New Era* (Gland, Switzerland: WWF International, 2016).

Index

Aboriginal peoples, 224–228
abstraction
    delaying gratification, 39–40
    language, 103–104
action-based semantics in dogs,
    178–180
HMS *Adventure*, 201
Adventure Bay, Tasmania, 201
agricultural runoff, domoic acid
    exposure from, 80–81, 90–91
agriculture, animal treatment and,
    245
Akeakamai (dolphin), 167
alternation task, 102
Alzheimer's disease, 73–74
amnestic shellfish poisoning,
    75–77
amphibians, 59–60
animal advocacy, 119–120, 240,
    251–252
Animal Liberation (Singer),
    242–243
animal welfare, 240–244, 247–248,
    251–252

Animal Welfare Act (1966), 241,
    243–244
A-not-B test, 43–46
anterior commissure, 206–207,
    220
anthropomorphization of animals,
    83, 103
anticipation, measuring, 143–144
aphasia, 71–72
Archer, Michael, 192–193, 199
Ashwell, Ken, 207–209, 212, 214,
    221–222
auditory cortex, 131, 134
auditory information processing,
    66, 129–131
Australia, marsupial dominance
    in, 191. *See also* Tasmanian
    devil; thylacines
autism, 112

backbone, development of, 57–59
Bailey, Col, 229–230
barking, 144–145
basal ganglia, 222

Bates, Stephen, 76
bats, 12–14, 134
"beat" frequency, 129
behavior reflecting mental state,
    16, 23
behaviorism: criticisms of the Dog
    Project, 140–142
Bennett, Craig, 138–139
Bentham, Jeremy, 240
Big Jack (dog), 35, 46, 149, 152,
    154–155
bigrams, 176
bilaterians, 55–56
birds
    A-not-B test, 43–44
    brain size and function, 66–67
    brain-to-body ratio, 62
    face processing, 174
    matching law, 153
    vocal learning and
        synchronization, 106–107
Black War (mid-1820s), 225, 228
Blackfish (movie), 119, 250
blobology, 138
blood oxygenation level dependent
    (BOLD) response, 33, 36
bottlenose dolphins, 122–123, 167
HMS *Bounty*, 201
bounty killing of thylacines,
    201–202, 227
Brain Ark, 259–260
brain imaging
    criticisms of, 138–140
    extrapolating experience across
        a range of animals, 245–246

See also diffusion tensor
    imaging; functional magnetic
    resonance imaging; magnetic
    resonance imaging
brain regions, 64–67, 69–70
    auditory processing in the
        brainstem, 129–131
    cerebral cortex, 130, 205,
        221–223
    hippocampus, 66–67, 89–91,
        93–99
    prefrontal cortex, 38–40, 223,
        232–233
    See also caudate nucleus;
        cerebellum; frontal lobe;
        temporal lobe; thalamus
brain size and structure
    action-based semantics,
        180–181
    amnestic shellfish poisoning,
        75–77
    A-not-B test as function of, 44
    as gauge for consciousness, 246
    brain-body size correlation,
        61–64
    connectomics studying, 71–72
    dolphin brains, 117–118
    domoic acid exposure, 84–85,
        90–91, 93–99
    evolution of the nervous system,
        54–56
    explaining differences in
        species, 6–7
    extrapolating experience across
        a range of animals, 245–246

gray matter, 67–70
  history of neuroscience, 51–54
  intelligence as a function of
    size, 63–64
  measuring brain size, 65–67
  movement as function of, 53–54
  nerve nets, 54–55
  purpose of size differentials in
    brain regions, 69–70
  species-specific functions,
    49–50
  swelling through *Pseudo-
    nitzschia* poisoning, 79
  themes of brain function, 50–51
  thylacine brain, 193–199
  understanding impulse control
    in dogs, 21–26
brain trauma
  domoic acid exposure, 84–85
  fMRI revealing brain activity,
    17
brainstem, auditory information
    processing in, 129–131
breeding programs: Tasmanian
    devils, 218–219
Buridan's ass, 153–154
bush, Tasmanian, 212–213

Cairo (military working dog), 1
Callie (dog), 2–11, 21, 35, 144,
    159–168, 173(fig.), 175
  expressing self, 180
  Food vs. Praise test, 144–145
  frontal lobe function in self-
    control, 47

Go-NoGo Task, 35–36
  learning to enter an MRI, 2–3
  name-face association, 168
  personality and temperament,
    11–12
  testing semantic knowledge,
    161–163
  word-object association test,
    159, 175
Campbell, Cameron, 199
canids, increasing brain study of,
    258
canine transmissible venereal
    tumor (CTVT), 218
carnivorous marsupials, 217–221.
    See also Tasmanian devil;
    thylacines
Casey, B.J., 40, 46
Cato (dog), 144–145
cats
  face recognition, 175
  frequency range, 127–128
  genetic engineering in, 254
  MRI training for, 5
  research demographics, 243
caudate nucleus
  basal ganglia and, 222
  facial recognition as positive
    emotion, 174
  measuring anticipation, 3–4,
    143–144
cats *(continued)*
  measuring regret in rats, 157
  praise-reward experiments,
    147–148

testing preferences and
   decision-making, 154–155
thylacine brain, 198
Caylin (dog), 163
central sulcus, 38, 122
cerebellum
   complexity of tasks, 112
   dolphin brain anatomy, 123
   thylacine brain, 198, 205
cerebral spinal fluid (CSF), 68
Chaser (dog), 160, 163, 166
children, Piagetian theory of
   cognitive development in,
   42–43
chimpanzees, 43–44, 63, 118, 160,
   246, 250, 252, 255
Church, George, 255
Churchill, Elias, 230–233
circuses, 250
Clements, Nicholas, 226, 228
Clever Dog Lab, University of
   Vienna, 170
clicks, dolphin, 127–128
climate change, 16, 200–201
Cnidaria, 54–55
cognitive dissonance, 245
cognitive processes
   auditory information
      processing, 66, 129–131
   delaying gratification, 39–40
   dolphin brain anatomy, 122
   effect of domoic acid exposure
      on memory, 90
   evolution of brain size, 61–64
   frontal lobe, 37–38

learning through echolocation
   study, 129–131
learning-by exclusion, 104–105
Piagetian theory of cognitive
   development, 42–43
pseudowords, 177–178
semantic representation,
   161–164
shared among humans and
   other animals, 103–104
specific brain regions, 64–67
visual information processing,
   66, 131, 222–223
cognitive psychology, 52
colonists' part in eradicating
   thylacines, 225–226
communication
   face processing, 174
   in dogs, 144–145, 161
   similarities in human-dog
      communication, 168
   See also language
comparative neurobiology, 7
computers, neuroscience and,
   52–53
conditioned responses, 51
conduction aphasia, 71–72
connectionism, 53
connectomics, 6, 71–72
consciousness
   brain injuries and, 71–73
   brain size as gauge for, 246
   self-awareness, 117–118,
      167–168, 245–246, 253
contextual training, 25

Cook, James, 200–201, 207, 219
Cook, Peter, 5, 20–21, 27, 45,
    83–90, 92–99, 108–115,
    116(fig.), 123, 146, 214, 237
  auditory processing in dolphins,
    130–133
  dolphin brain scans, 120–125
  domoic acid exposure study,
    102–103
  Food vs. Praise experiment,
    145–148, 155
  Go-NoGo tasks for impulse
    control, 34–41
  pinniped lab, 83–90
  sea lion brain imaging, 92–99
  sea lion bran data, 5
  studying rhythm in sea lions,
    106–113, 115–116
  thylacine brain scan, 197–198
Cooper, Matthew, 248–249, 252
corpus callosum
  dolphin brain anatomy, 125
  DTI scans in sea lions, 97
  function of, 71–72
  Tasmanian devil's lack of, 220
  thylacine brain, 198, 206–207
cortex, cerebral
  auditory information
    processing, 130
  thylacine and Tasmanian devil
    brains, 221–223
  thylacine brain, 205
cosmology: appearance of animals,
    54–55
coyotes, 258

crepuscular animals, 186–188
Cretaceous-Tertiary event, 61
CRISPR/Cas9, 254
crocodiles, 60
crown mammals, 190, 220
cruciate sulcus, 38, 122
Cruelty to Animals Act (1849),
    241
cryptids, 190
curiosity in dogs, 2
custody issues, 248–249
cynodonts, 60–61

D'Amico, Vicki, 151
dancing, 109–110
Darwin, Charles, 105–106
Darwinian selection. *See* evolution
Dasyuromorphia, 217
death: gross anatomy lab, 236–238
decisionmaking skills
  disappointment and regret
    driving, 155–156
  early brain research, 138
  evolution of the brain, 56–57
  examining future outcomes, 59
  Food vs. Praise in dogs,
    150–154
delayed alternation test, 88–90
delayed gratification, experiments
    in, 39–41
Dennison, Sophie, 86
Descartes, René, 243
The Descent of Man (Darwin),
    106
devil facial tumor disease

(DFTD), 218

diffusion tensor imaging (DTI)
auditory information processing in dolphins, 132
domoic acid exposure in sea lions, 93–99
expanding research on, 257–258
mapping dolphin brains, 118–119
mapping hippocampal connections in sea lions, 93–99
thylacine brain, 198, 206–207

digital seeds, 99, 131–134

Dilks, Danny, 172

dinosaurs, 60–61

diprotodont (giant marsupial), 224

disappointment driving learning, 155–156

disease
(Tasmanian) devil facial tumor disease, 218
domoic acid exposure, 75–77
gross anatomy, 236–238

dissection, 236–239

DNA editing, 254

dog face area (DFA), 174

dog lab, 238–240

Dog Project
animal welfare and animal rights, 235–240, 248
comparing human and dog brains, 50
criticisms of, 140–142

dogs recognizing human faces, 171–172
Food vs. Praise experiment, 142
improving the methodology, 139–140
MRI training for dogs, 139–140
principles of research, 235
refining the research, 47–48
research goals of, 4–6, 22

dogs
experiencing anticipation and regret, 157
face processing, 170–175
Food vs. Praise, 144–155
frequency range, 127–128
language understanding, 166
research demographics, 243
sea lion brains and dog brains, 125–126
sheep killings in Tasmania, 227–228
testing preferences, 150–153
thylacines' resemblance to, 189–190
understanding emotional states in humans, 170–175
word-object associations, 160, 164–167, 175–178

dolphins
brain anatomy, 117–118, 122–126, 133–135
brain imaging, 8
connections with terrestrial mammals, 133–136
echolocation, 126–129

encephalization quotient, 117
language comprehension, 167
measuring sentience, 242
domoic acid exposure, 76–77,
80–81, 84–85, 90–91, 93–99,
102–103
Dostoyevsky, Fyodor, 91
The Dress (Internet phenomenon),
136
dunnarts, 217

Earth, Wind and Fire, 111
echolocation
acuity and speed in dolphins,
128–129
animals utilizing, 134–135
as active process, 132
dolphin brain anatomy,
126–127
quantifying a subjective
experience, 13–14
range of dolphin sounds, 118
similarities between humans
and dolphins, 136
temporal auditory tract,
133–134
testing subjective experience in
dolphins, 129–131
Eddie (dog), 10, 36, 46
Edison, Thomas, 51
education: gross-anatomy and dog
lab, 236–239
egg-laying animals, 60
El Niño, 79–81, 85, 97–98
electrochemical processes in the

brain, 51–52, 72–73
elephant seals, 257–258
elephants, 250
emotions in animals
arbitrary one-off studies,
137–138
dogs understanding emotional
states in humans, 170–175
questioning the existence of,
16–17
response to anticipation of
pleasure, 4–5
sharing the experience of, 15
encephalization quotient (EQ),
63, 117
entrainment, 106, 109
environmental niche, interpreting
brain structure in the
context of, 212
epilepsy, 91, 99
ethics
animal welfare, 243–244, 251
Marine Mammal Protection
Act, 130–131
pain studies, 135
evolution
appearance and evolution of
fish, 59–61
evolution *(continued)*
auditory information
processing, 131
development of the backbone,
57–59
marsupials and placental
mammals, 190–191

measuring brain size, 64–67
of human language, 166
of mammalian brains, 220–221
of marsupials, 190–191, 224
of the nervous system, 54–57
role of language in human
  evolution, 103
expected utility theory, 155
extinction of species
  Aboriginal populations, 226
  genetic engineering, 253–255
  Homo sapiens, 253, 255
  insurance population for
    endangered species, 219
  interconnectedness of species
    survival, 259–260
  mass extinction events, 60–61
  prompting scientific research,
    192–193
  thylacine, 189–190

face processing, animals' ability
  for, 170–175
facial recognition, 173–175
Fagan, Jessa, 41
Farm Animal Welfare Council
  (1979), 242
fear, mapping in the brain, 174
Federal Brain Research through
  Advancing Innovative
  Neurotechnologies (BRAIN)
  initiative, 259
fish
  development of the backbone,
    57–59

evolutionary path, 59–61
genetic engineering in, 254
measuring sentience, 242
Monterey Bay Aquarium
  research, 81–83
Pseudo-nitzschia poisoning,
  80–81
Fleming, Arthur, 202
flying, explaining the response to,
  13–14
food
  animals as, 245
  Food vs. Praise experiment,
    142, 148–155
  foraging styles, 257–258
  sea lion foraging tests, 87–88
  thylacine foraging habits,
    214–215
food poisoning, 75–77
Food vs. Praise experiment, 142,
  148–155
foraging test, 87–89
Frank-Starling law, 237
free will, 151–153
frontal lobe
  A-not-B test, 46
  dolphin anatomy, 122
  Go-NoGo task, 46
  human versus dog brains, 47,
    49–50
  impulse control, 148–149
  NoGo trial activating, 37–38
  perseveration error, 43
  Tasmanian devil brain, 220
functional magnetic resonance

imaging (fMRI)
benefits of, 137–138
BOLD response, 33
brain coding of visual scenes,
162–163
brain trauma findings, 17
face processing in dogs, 171–172
finger-tapping experiments
studying rhythm, 111–112
Go-NoGo task, 35–36
limited resolution of, 50
measuring anticipation,
143–144
praise-reward experiments,
147–148
questioning the results of,
138–139
shallow level of language
processing in dogs, 178–179
word-object associations in
dogs, 175–176
fusiform face area (FFA), 170

gadolinium, 94
Gallant, Jack, 17, 162–164,
169
Gallup, Gordon, 117–118
G-Dock (sea lion), 102–103
genetic engineering, 254–255
glial cells, 68
global properties of objects, 165
glutamate, 76
goal-directed actions, 73
Go-NoGo task
analyzing self-control, 38–39

brain activation areas, 40–41
first brain imaging studies, 40
fMRI scans, 35–37
impulse control, 21–23
MRI scans, 28–35
performing a behavior during
an MRI scan, 19–21
questioning the Dog Project
methodology, 141
testing diverse dogs' impulse
control, 33–35
gray matter
brain composition, 67–70
dolphin brain anatomy, 123
thylacine brain, 198
gross anatomy, 236–238
Guiler, Eric, 200, 202–203,
227–228
Gulland, Frances, 77–78, 80–81,
85, 90, 102, 212

hagfish, 57–58
hand signals, 3–4, 37, 104, 108,
143–144
Harari, Yuval, 253
harbor seals, 257–258
Hare, Brian, 23–24
hearing, dolphins' mechanism for,
128–129
Herculano-Houzel, Suzana, 63–64
Herzog, Hal, 245
hippocampus
as site of seizures, 91
domoic acid exposure and
memory, 90

mapping hippocampal connections through DTI in sea lions, 93–99

size-function relationship in birds, 66–67

testing sea lions, 89

hippopotamus, echolocation in, 134–135

Hobart Zoo, Tasmania, 185–187

Hogg, Carolyn, 219

Homo hominis, 255

Hoover (harbor seal), 257–258

Hrdlička, Aleš, 194–195

Huber, Ludwig, 170

human development: Piagetian theory of cognitive development, 42

human genome sequencing project, 254

The Idiot (Dostoyevsky), 91

imagination
experiencing regret, 156
fMRI revealing brain activity despite trauma, 17

impulse control, 21–26, 34, 147–148

individual behavior versus species behavior, 46–48

inferior colliculus, 66, 130–132

Ingleby, Sandy, 213

input-output analogy of brain function, 53–54

Institute for Marine Sciences, 82–84

Institutional Animal Care and Use Committees (IACUCs), 241, 252

insurance population of Tasmanian devils, 218–219

intelligence, brain size and, 63–64

interior frontal gyrus (IFG), 40

internal versus external perspectives, 14–15

International Thylacine Specimen Database (ITSD), 193

International Union for the Conservation of Nature and Natural Resources (IUCN), 218

Jackendoff, Ray, 166

James, William, 15

jellyfish, 54–55

Jerison, Harry, 61–62, 68–70

Jettyhorn (sea lion), 86–89

Joey (dog), 248–249, 252

Kady (dog), 11
A-not-B test, 46
Food vs. Praise test, 142, 146, 148–150, 152
Go-NoGo Task, 23–28, 31–32, 31(fig.), 33–34
marshmallow test, 41–42
mental inertia, 22–23

Keen, Cindy, 35, 152, 155

Kimmela Center for Animal Advocacy, 119–120

King, Patricia, 22–28, 31, 31(fig.),
    33–34, 41, 146
Krebs, John, 67

lampreys, 57–58
Langan, Esther, 193–196
language
    abilities in cetaceans, 167
    action-based semantics,
        178–180
    as differentiating feature of
        humans, 103–105
    conducting aphasia, 71–72
    fMRI research on semantic
        representations, 162–164
    human semantic space, 169–171
    pidgin languages, 166
    semantic representation,
        161–164
    sharing experiences through,
        14–15
    sign language with sea lions,
        82
    subjective experience and,
        157–158
    word-object association in
        dogs and humans, 159–161,
        164–169, 176–178
law of effect, 51
learning, dogs' curiosity for, 2
learning-by exclusion, 104–105
left-right coordination, 56, 71–72,
    151–152
Libby (dog), 11–12, 12(fig.), 18–22,
    25–26, 34–35, 142

Lilly, John, 167
Little Hoot (sea lion), 92–99
Living Planet Report, 259
local properties of objects, 165
logical symmetry, 105
Lunde, Darrin, 193–196

magnetic resonance imaging
        (MRI)
    acclimating the dogs to the
        noise of, 30–31
    advancing technology, 252–253
    comparing thylacine and
        Tasmanian devil brains,
        221–223
    complexity and mechanics of,
        28–30
    dogs' right to refuse, 244–245
    dogs' tendency not to move
        inside, 34–35
    dolphin brain anatomy, 117–
        118, 121–125
    DTI data and, 94
    flex coil, 197–198
    Go-NoGo Task, 28, 35–36
magnetic resonance imaging
        (MRI) *(continued)*
    optimizing scan speed and
        gradient power, 198–199
    praise-reward test, 146–150
    questioning results of, 138–
        139
    science of connectomics, 6
    sea lion brains, 5, 86–88
    studying rhythm, 112

Tasmanian devil brain, 219–221
thylacine brain, 195–198,
203–209, 213–217
training other animals for, 5
word-object associations in
dogs, 175–178
Magritte, René, 161
mammals, encephalization
quotient in, 63
manatees, 257–258
Marconi, Guglielmo, 51
Marine Mammal Center,
Sausalito, California,
77–81, 89–90, 92–93, 97–98,
101–102, 257–258
Marine Mammal Protection Act,
130–131
Marino, Lori, 117–122, 132
marshmallow test, 39–42, 157
marsupial lion, 224
marsupials
brain anatomy, 207–208
evolution of placental mammals
and, 190–191
settlers' contribution to the
extinction of, 201–202
See also thylacines
matching law, 153
medial geniculate nucleus,
130–131
medical school: gross-anatomy
and dog lab, 236–239
melon (on dolphins), 127
memory
animals' recognition of

individual humans, 114
brain connectivity and, 73–74
domoic acid exposure and,
88–90
role of the hippocampus, 67
sea lions, 84–85, 89
word-object association in dogs,
178
mental experience
action-based worldview,
179–181
observing behavior, 16
other animals experiencing,
15
mental representations of words,
161–164
metatheria, 190–191
mice, 246–247, 258–259
Mike (leopard), 185–186
military working dogs, 1
Miller, Karla, 94
Mills, Daniel, 165
Mischel, Walter, 39, 46
Molaison, Henry, 67
monkey lips/dorsal bursae
(MLDP), 127
monotremes, 190, 207
Monterey Bay Aquarium,
Monterey, California, 79–82
moral issues
animal welfare legislation,
247–248
the power of science to shape
moral intuitions, 253
mosaic evolution, 65–67

motivation, questioning dogs',
141–142
motor cortex and motor function
basal ganglia and, 222
comparing thylacine and
Tasmanian devil brains,
222–223
rhythm experiments, 111–112
movement
as function of the brain, 56–57
cerebellum processing, 112
effect on fMRI scans, 36
explaining the experiences of,
15
functions of brain regions,
65–67
synchronizing auditory systems
and, 113–116
MRI simulator, 3, 9, 18–19, 141
Munroe, Randall, 136
Murry, Trisha, 248
music
origins in non-human species,
105–107
teaching synchronization in sea
lions, 108–113
mussels, 75–76
myelin, 68–70

Nagel, Thomas, 12–13, 52–53,
118, 129
names, animals' understanding of,
167–168
National Institutes of Health
(NIH), 250

National Zoo, Washington, D.C.,
194, 195(fig.)
natural selection, 253
Neanderthals, 255
nematocysts, 55
neoteny, 83
nerve nets, 54–55
nervous systems, 54–56
neural-net models, 53
neurons
encephalization quotient and,
63–64
gray matter, 67–70
neuroscience
expanding research on
nonhuman species, 258–259
shifting paradigms of study,
70–71
subjective experiences and,
12–13
neurotransmitters, 76
New York Times, 141, 248
Newton (dog), 1
Non-Human Rights Project, 252
novelty triggering survival
processes, 178
numbats, 217
Nuremberg Trials, 244

object permanence, 42
objective experiences, 13–14
Office of Naval Research, 4
Ohana (dog), 10–11, 36, 149(fig.),
154
olfactory bulb, 65–66, 205–206,

214
one-off studies, 137–138
operant conditioning, 51, 88
optogenetics, 253
orbitofrontal cortex, 157
"The Origin and Function of
    Music" (Spencer), 106
osmiridium mining, 230
Owen, Adrian, 17
Ozzie (dog), 149

pain and suffering, 235, 242–244
pantropical dolphins, 133–135
parsimony, principle of, 107
Patel, Aniruddh, 106–107
Pavlov, Ivan, 51
Pavlovian experiments, 141
Pearce, Claire, 11, 18–20, 34
Pearl (dog), 10–11, 45–46,
    148–150, 149(fig.), 151–152,
    154
perseveration error, 43
personality affecting MRI trials,
    34
phase shifting, rhythm and tempo,
    115–116
Phoenix (dolphin), 167
phonation, 127
phrenology, 138
phytoplankton, 77–81
Piaget, Jean, 42
pictograms: language
    comprehension in animals,
    104–105
pidgin languages, 166

pigs, MRI training for, 5
Pilley, John, 160
pinnipeds, 82–86
piriform cortex, 205
placental mammals, 191, 222
pointing tasks, 23–27, 33–34
Poldrack, Russ, 139–140
Popov, Alexander, 51
praise, analyzing in dogs, 142,
    145–146
praise vs. food, 144–155
praise-reward test, 146–150
preferences in dogs, 150–153, 169
prefrontal cortex and activity
    complexity in thylacine brains,
        223
    Go-NoGo task, 38–39
    inferior frontal gyrus, 40
    thylacine brains, 232–233
primates
    A-not-B test, 43–44
    banning chimpanzee research,
        250
    language ability, 160–161
    measuring sentience, 241–
        242
probabilistic tract tracing, 99
proper mass, principle of, 61–62
property, animals as, 248–249
Pseudo-nitzschia phytoplankton,
    77–81
pseudowords, 177–178
psychology, history of
    neuroscience and, 51–52
pulvinar, 130–131

putamen, 222

qualia (subjective experience),
   135–136
quolls, 201, 217

raccoon brain, 196–197
rape of Aboriginal girls and
   women, 226
rate-flexibility, 109
rats
   BRAIN initiative, 258–259
   matching law, 153
   neural basis for regret in,
      156–157
   sentience order, 242
Redish, David, 156–157
reduction (animal research rubric),
   242
refinement (animal research
   rubric), 242
reflexive behavior, 51
regret, decisionmaking and,
   156–158
Reichmuth, Colleen, 82–84, 88,
   103–104, 112–113, 115
Reid, Alison, 184–187
Reiss, Diana, 117–118
replacement (animal research
   rubric), 241–242
reproduction, drive for, 56–57
reverse inference, 140–142
rewards, experiments in, 142
rhythm in animals, 106–115
Rico (dog), 160

Ringling Bros. and Barnum &
   Baily Circus, 250
Rio (sea lion), 104–105, 113
Rocky (sea lion), 104, 166
Ronan (sea lion), 101–116,
   116(fig.)
Rouse, Andrew, 113–116
rubber-hand illusion, 246–247
Rudi, Patti, 149

sauropsids, 60
Schusterman, Ron, 82, 104, 166
sea lions
   adopting for study, 102–103
   compensating for hippocampus
      damage, 98–99
   delayed alternation test, 88–90
   dolphin brains and, 124
   effect of domoic acid and
      hippocampus damage, 92
   Institute for Marine Sciences
      mission, 82–83
   language acquisition, 104–105
   mapping hippocampal
      connections through DTI,
      93–99
   pidgin-level language
      understanding, 166
   Pseudo-nitzschia poisoning in,
      77–81
   recognition of individual
      humans, 114
   stranding behavior, 5
   studying domoic acid toxicity
      in, 84–86

studying rhythm in, 106–107
teaching synchronization,
    108–113
Sea World, 250
seals, 113, 257–258
seizures, 76
Sejnowski, Terrence, 68–70
self, sense of, 73–74
self-awareness, 117–118, 167–168,
    223, 245–246, 253
self-control, analyzing, 37–42,
    46–47
self-determination in animals,
    236, 244
semantic knowledge and semantic
    space
  action-based, 178–181
  dogs' understanding of human
    emotions, 171
  face and emotion recognition in
    dogs, 174
  function of names, 167–168
  in primates, 160–161
  pidgin languages, 166–167
  studying in humans, 161–163
  testing in dogs, 161–166
  word-object associations in
    dogs, 175–178
semantics, defining, 169–170
sensory perception
  hand signal response, 3–4
  rubber-hand illusion, 246–247
sensory-motor coordination:
    A-not-B test for children, 43
sentience, 141, 223–224, 241–242,
    246, 250, 253, 258
service dogs, 9–11, 23
settlers in Tasmania, 200–201
sharing experiences, 14–15
sheep killers, thylacine as, 201–
    203, 215, 226–228
shellfish poisoning, 75–81
sight, echolocation and, 133–135
Singer, Peter, 242–243
Sleightholme, Stephen, 193, 199,
    208, 229
Smithsonian Institution, 193–196
Snowball (cockatoo), 107
social reward, 143–144
sonar (sound navigation and
    ranging), 126–127
sound production
  in dolphins, 127–129
  vocal flexibility in pinnipeds,
    257–258
spatial memory, 67
species behavior versus individual
    behavior, 46–48
speciesism, 4–5
Spencer, Herbert, 106
Spivak, Mark, 2–3, 28, 45, 139,
    151,
state transitions, 154–155
status epilipticus seizures, 78–79
strokes, 71–72, 203
subcortical structures, 222
subjective experience
  brain imaging, 7–8
  dolphin brains, 118
  explaining and quantifying,

12–14
  incentive for studying, 239–240
  qualia arguments, 135–136
suffering and pain, 235, 242–246
supercomputing, 97
superior colliculus, 66, 130
superior temporal lobe, 177–178
surface-area rule of brain size,
  62–63
survival, drive for
  evolution of the modern brain,
    56–57
  increasing brain complexity, 59
  novelty triggering, 178
symbolic representation, 15
synapses, 68–70
synchronization in animals,
  106–107
syntax, animals learning, 166–167

T1 quantity (MRI), 198, 207
T2 quantity (MRI), 198
Taronga Conservation Society, 218
Tasman, Abel, 200
Tasmanian devil
  comparing thylacine brains
    with, 221–223
  MRI scans, 219–221
  recognition of animal sentience,
    250–251
  threats to the survival of,
    217–218
Tasmanian tiger. *See* thylacines
Tasmanian Wilderness World
  Heritage Area, 229–233

taxidermy, 185
temporal lobe
  auditory information
    processing, 131, 133–134
  dolphin brain anatomy, 125
  facial recognition in dogs,
    173–174
  hippocampus and, 66–67
terrestrial animals' similarities to
  dolphins, 133–136
tetrapods, 59–60
thalamus
  auditory information
    processing, 130–132
  effect of domoic acid exposure,
    99
  thylacine and Tasmanian devil
    brains, 221–222
  thylacine brain, 208
thermoregulation, 60–61
Thorndike, Edward, 51
three R's, 241–243
The Thylacine Museum (website),
  199–200
thylacines, 8, 195(fig.)
  as sheep killers, 201–203,
    227–228
  brain scans, 190–199, 208–209,
    213–217, 216(fig.)
  causes of the decline in the
    population, 199–203
  colonists' part in eradicating,
    225–226
  comparing Tasmanian devil
    brains with, 221–223

data base for, 193

decline and extinction,
199–203

foraging habits, 214–215

Hobart Zoo, 183–184, 188–189

looking for proof of existence,
211–213, 229–233

shrinkage in the stored brain,
203–205

similarities to dogs, 189–190

speed of, 216–217

Tasmanian devil brain and, 220

temperament, 186–187

tinkering, era of, 253–255

toys, experiments with praise,
145–146

training

A-not-B test, 44–46

function of animals' names, 168

Go-NoGo task, 20–21

language acquisition in
animals, 103–105

MRI scans of dogs, 2–3

MRI simulator, 18–21

pointing tasks, 23–26

positive reinforcement, 235

sea lions, 82–83, 87–88

starting with a natural
behavior, 108–109

word-object associations,
159–160, 165–166

transitivity, language and,
105

traumatic brain injury, 72

Travis, Shannon, 248

tricks, teaching dogs, 165–166

Truffles (dog), 149

Tug (dog), 41–42

Turner, Ted, 192

Udell, Monique, 24

University of California at Santa
Cruz, 83–84

USDA Predator Facility, Utah,
258

utilitarianism, animal welfare and,
240–241, 251

Van Diemen's Land Company, 227

vegan issues, 4–5

Velcro (dog), 148–150, 152,
154–155

vertebrates, development of the
brain in, 57–59

Vienna, University of, 170

visual cortex

dolphins processing auditory
information, 134

measuring activity in, 17–18

visual information processing,
66

comparing thylacine and
Tasmanian devil brains,
222–223

dolphin brain anatomy and, 131

vivisection, 238–249

vocal mimicry, 106–107, 112–113

"Voodoo Correlations in Social
Neuroscience," 139

voxels (volume elements), 36
Vul, Ed, 138–139

Walken, Christopher, 110
warm-blooded animals, 60–61
whales, echolocation in, 127
"What Is It Like to Be a Bat?"
    (Nagel), 12–13
white matter, 68–70, 93, 95–96,
    98–99
  dolphin brain anatomy, 123,
    126(fig.)
  thylacine brain, 198

white rhinos, 251
wildlife advocacy, 251–252
Wise, Steven, 252
word-action associations, 178–179
word-object associations, 160–164,
    175–178
World Wildlife Fund, 259–260

xkcd, 136

Zen (dog), 9–12, 19, 46
Zhang, Kechen, 68–70

Helen Berns

Gregory Berns, MD, PhD, is a professor of psychology at Emory University, where he directs the Center for Neuropolicy and Facility for Education and Research in Neuroscience. He is the author of several books, including the *New York Times* bestseller *How Dogs Love Us*. He lives in Atlanta with his wife and too many dogs.